Lectures on Quantum Mechanics and Relativistic Field Theory

P. A. M. Dirac

Notes by
K.K.Gupta and George Sudershan

Martino Publishing
Mansfield Centre, CT
2012

Martino Publishing
P.O. Box 373,
Mansfield Centre, CT 06250 USA

www.martinopublishing.com

ISBN 978-1-61427-334-9

Cover design by T. Matarazzo

Printed in the United States of America On 100% Acid-Free Paper

Lectures on Quantum Mechanics and Relativistic Field Theory

P. A. M. Dirac

Notes by
K.K.Gupta and George Sudershan

STEVENS & CO.
1841 BROADWAY, NEW YORK 23, New York

1955

CONTENTS

Lecture 1

Newtonian mechanics has been most successful in describing the motion of systems consisting of big, ponderable, objects. For the description of atoms one needs a new mechanics which is variously called Wave Mechanics, Matrix Mechanics or Quantum Mechanics. These different names all describe essentially the same theory. In the present development of Quantum Mechanics the method followed will be deductive - a number of assumptions will be laid down and the conclusions and deductions will then be worked out from them.

The concept of 'small' objects for whose description quantum mechanics is necessary can be given an absolute meaning. The fundamental constants of nature e, m and h allow one to construct fundamental units which can serve as a criterion for 'bigness' and 'smallness' in the absolute sense. An object is big (in the absolute sense) if its mass and dimensions are much larger than those provided by these units; otherwise it is small. The main characteristic of a small object is that the inevitable disturbanace produced in it on making a physical observation of it is not negligible. It cannot be cut down by the experimenter being more careful and using improved technique in making the observation. This results in an indeterminacy in the results of observation on small systems: the results of two identical observations on the same system will not in general be the same. Only statistical results are of significance for such systems.

In classical mechanics the state of a given system can be defined by giving the positions and momenta of the particles constituting it at a given time. In quantum mechanics one cannot specify a state in the same way since one cannot determine all the dynamical variables of the system at a given time. A state in quantum mechanics is determined by specifying the values of as many dynamical variables as the theory allows one to specify simultaneously, at a given time.

The ugly feature of indeterminacy of the results of observation on a system is compensated by the principle of super-position of states which will now be introduced. Consider the set of all the states of a dynamical system. The principle of super-position allows one to superpose any two states to get other states belonging to the set. This property of the states makes it convenient to represent them by vector in a vector space. These vectors will be called kets and denoted by the symbol $|\rangle$. If it is desired to distinguish a ket from another a label can be inserted in it thus: $|A\rangle$. The space of kets is in general infinite dimensional. The ket vectors satisfy the following assumptions:

(i) Kets are complex, i.e., a ket can be multiplied by a complex number to obtain a quantity of the same nature, i.e., another ket.

(ii) There is a one to one correspondence between the states of a dynamical system and the directions of the ket vectors. Thus $|A\rangle$ and $c|A\rangle$, where c is a complex number, correspond to the same state. This gives a difference between the quantum mechanical superposition principle and the superposition principle which sometimes exists in classical mechanics. The superposition of a mode of vibration of a string on itself gives, in classical physics, a new mode with twice the amplitude and four times the energy of the initial mode of vibration. In quantum mechanics the superposition of a state on itself gives rise to the same state. Also the zero vector in quantum mechanics corresponds to no state at all, whereas the zero vector in the example of vibrating string corresponds to a string at rest.

Consider two states $|A\rangle$ and $|B\rangle$ for which the results of measurement of a dynamical variable ξ, say, are respectively a and b. What is the result of measurement of ξ on the superposed state

$$c_1 |A\rangle + c_2 |B\rangle \quad ?$$

The system is now partly in state $|A\rangle$ and partly in state $|B\rangle$. The act of measurement of ξ will force the system into one or other of these states and the result obtained will be a or b, with probabilities depending on the relative weightages of the states $|A\rangle$ and $|B\rangle$ in the superposed state. This is the manner in which the principle of superposition can be understood physically with the help of the principle of indeterminacy.

The ket

$$c_1 |A\rangle + c_2 |B\rangle$$

obtained by superposing the kets $|A\rangle$ and $|B\rangle$ furnishes a two-fold infinity of states, since only the direction of the new ket, and hence only the complex number c_1/c_2, matters in so far as the characterisation of the state is concerned. If one considered only real kets, one would arrive at a one-fold infinity of states in this manner; this would not correspond to reality as the following example shows. Consider a beam of light of given frequency moving in a definite direction. It consists of photons of definite energy and momentum, which may be in various possible states of polarization. Now there are only two independent states of polarization, which may be taken as, for example, (1) the 2 states of linear polarization in perpendicular directions; or (2) the

2 states of circular polarization in opposite directions. The general superposed state would be a state of elliptic polarization. Since an ellipse requires two parameters for its characterization one sees that a two fold infinity of states is actually obtained when two independent states of polarization of a photon are superposed. Hence it is that one must employ complex kets to represent states of a physical system.

It can be shown that for a given vector space there always exists another vector space called the dual vector space, such that one can form a scalar product of two vectors, one from each space. The vectors of the space dual to that of ket vectors will be called bra vectors or bras and will be denoted by $\langle\,|$. The scalar product of ket $|A\rangle$ and bra $\langle B|$ will be denoted by $\langle B|A\rangle$. This is a complex number.

One will find in the following that completed brackets are always numbers and uncompleted ones vectors.

It is assumed that

(iii) There is a one to one correspondence between kets and bras. The bra corresponding to ket $|A\rangle$ will be denoted by $\langle A|$.

(iv) The above correspondence is antilinear. That is

$$|A\rangle + |B\rangle \longrightarrow \langle A| + \langle B| ,$$

$$c|A\rangle \longrightarrow \bar{c}\,\langle A| .$$

(v) $$\langle A|B\rangle = \overline{\langle B|A\rangle} .$$

(vi) $$\langle A|A\rangle > 0 \quad \text{for} \quad |A\rangle \neq 0 .$$

The bra and ket vectors $\langle A|$ and $|A\rangle$ can be looked upon as

conjugate imaginary of each other.

Two kets are said to be orthogonal if the scalar product of one with the bra corresponding to the other is zero. Similarly one defines the orthogonality of two bras and of a bra and a ket.

A ket $|A\rangle$ is said to be normalized if

$$\langle A|A\rangle = 1.$$

To a state there correspond all the kets having a given direction. One can make the correspondence more definite by assigning to a state only normalized kets. There will still be an arbitrariness in the kets, since they can be multiplied by a factor $e^{i\gamma}$, γ real, without destroying the normalization. $e^{i\gamma}$ is called a phase factor.

From assumptions (ii) and (iii) it is seen that to every state there also corresponds a direction in the space of bra vectors. Indeed, there is a complete symmetry between bra and ket vectors and one might as well have started with bras for the characterization of states.

Lecture 2

We now introduce linear operators acting on the kets. Let α be a linear operator; acting on a ket $|A\rangle$ it gives another ket depending linearly on $|A\rangle$ which is written as the product $\alpha|A\rangle$. Sums and products of linear operators can be directly defined:

$$(\alpha+\beta)\,|A\rangle = \alpha\,|A\rangle + \beta|A\rangle$$
$$\alpha\beta\,|A\rangle = \alpha\,(\beta|A\rangle)$$

Thus an algebra of these operators can be built up. In general the commutative law of multiplication is not satisfied:

$$\beta\alpha\,|A\rangle \neq \alpha\beta\,|A\rangle\,,$$
$$\beta\alpha \neq \alpha\beta\,.$$

If for two operators α, β $\alpha\beta = \beta\alpha$ we say that α and β commute.

Any number k can multiply a ket $|A\rangle$ giving a ket $k|A\rangle$ depending linearly on $|A\rangle$. Hence a number is a linear operator. It has the property of commuting with all linear operators.

So far linear operators have been defined to act on kets. We can make a linear operator α act on a bra $\langle B|$ to give a result written as the product $\langle B|\alpha$, by assuming

$$(\langle B|\alpha)\,|A\rangle = \langle B|\,(\alpha|A\rangle)\,.$$

With this definition of $\langle B|\alpha$ the triple product can be taken to be associative and written simply as $\langle B|\alpha|A\rangle$; one can form products like $\langle P|\alpha\beta\gamma|Q\rangle$ the multiplication being always associative.

The physical meaning to be attached to linear operators is that they correspond to dynamical variables, like position, velocity, of a particle and field variables in a field theory. Thus the dynamical variables of quantum mechanics obey a non-commutative algebra.

Given a bra $\langle B|$ and a ket $|A\rangle$ we can form a product $|A\rangle\langle B|$ which one can interpret to be a linear operator by demanding the associative law to be valid:

$$\langle Q|\,(|A\rangle\langle B|) = (\langle Q|A\rangle)\,\langle B| .$$

This shows that $|A\rangle\langle B|$ acting on a bra $\langle Q|$ gives another bra which depends linearly on it and is hence a linear operator.

The products we have enumerated include all combinations to which a meaning can be attached. Other combinations like $\alpha\langle P|$ are meaningless.

In general, linear operators are to be considered complex. To make a linear operator correspond to a real dynamical variable or to a real function of dynamical variables, we have to impose some restrictions on it. Given any linear operator α there is a linear operator $\bar{\alpha}$ which is the complex conjugate of α which is defined by

$$\langle A|\bar{\alpha}|B\rangle = \overline{\langle B|\alpha|A\rangle} .$$

Then $\bar{\bar{\alpha}} = \alpha$. What we have called the complex conjugate of an operator is called the adjoint operator by mathematicians. In view of the physical interpretation, "complex conjugate" is to be preferred in quantum theory.

The conjugate so defined obeys certain rules. Put

$$\alpha|A\rangle = |P\rangle$$

Since $\langle A|\bar{\alpha}|B\rangle = \overline{\langle B|\alpha|A\rangle} = \overline{\langle B|P\rangle} = \langle P|B\rangle$

for every ket $|B\rangle$,

$$\langle A|\bar{\alpha} = \langle P| = \text{conjugate of } \alpha|A\rangle .$$

Take two linear operators α, β and form $\alpha\beta|A\rangle$. Its conjugate is $\langle A|\overline{\alpha\beta}$. But conjugate of $\alpha\beta|A\rangle$ is conjugate of $\alpha(\beta|A\rangle)$ and applying the relation just derived

$$\text{conj. } \alpha(\beta|A\rangle) = (\text{conj } \beta|A\rangle)\bar{\alpha} = \langle A|\bar{\beta}\bar{\alpha}$$

and hence

$$\overline{\alpha\beta} = \bar{\beta}\bar{\alpha}$$

Similarly, if we have more than two factors $\alpha, \beta, \gamma, \dots$

$$\overline{\alpha\beta\gamma\dots} = \dots\bar{\gamma}\bar{\beta}\bar{\alpha}.$$

All these rules can be summarised in the single comprehensive rule which says that to get the conjugate of any product, one has to take the product of the conjugates of the factors in reverse order. Take $|A\rangle\langle B|$ for example

$$\langle P|\overline{(|A\rangle\langle B|)}|Q\rangle = \overline{\langle Q|A\rangle\langle B|P\rangle}$$

$$= \langle A|Q\rangle\langle P|B\rangle = \langle P|B\rangle\langle A|Q\rangle$$

since $\langle A|Q\rangle$ and $\langle P|B\rangle$ are numbers and hence commute.

Hence

$$\overline{|A\rangle\langle B|} = |B\rangle\langle A| .$$

Another illustration is the old result

$$\langle A|B\rangle = \overline{\langle B|A\rangle} .$$

To introduce coordinates into the space of ket vectors, consider the basic kets $|\gamma\rangle$ satisfying the following conditions:

(i) They form a complete set; any ket $|P\rangle$ can be expressed in the form

$$|P\rangle = \sum_r a_r \, |r\rangle .$$

(ii) They are orthogonal and normalised:

$$\langle r|s\rangle = \delta_{rs} .$$

(iii) They are linearly independent:

$$\text{if } \sum_r a_r \, |r\rangle = 0 , \qquad \text{then } a_r = 0 .$$

(iv) $$\sum_r |r\rangle\langle r| = 1 ,$$

where 1 is the unit linear operator.

These conditions are not all independent. If (i) and (ii) are assumed, the other two can be derived. Thus, assuming (i) and (ii), if we have $\sum_r a_r \, |r\rangle = 0$ we multiply by $\langle s|$ and get

$$\sum_r a_r \langle s|r\rangle = 0 .$$

Using (ii) this gives $a_s = 0$, so (iii) is proved.

Again, take $\sum_r |r\rangle\langle r|$ and multiply by $|s\rangle$ getting

$$\sum_r |r\rangle\langle r|s\rangle = \sum_r |r\rangle \delta_{rs} = |s\rangle ,$$

so that $\sum_r |r\rangle\langle r|$ applied to $|s\rangle$ gives $|s\rangle$ itself. Since the $|s\rangle$ form a complete set and span the whole space of kets, $\sum_r |r\rangle\langle r|$ acting on any ket would give the same ket so that

$$\sum_r |r\rangle\langle r| = 1 ,$$

which is condition (iv).

Alternatively, assuming (iii) and (iv), one can derive (i) and (ii).

The equation $|P\rangle = \sum_r |r\rangle\langle r| P\rangle$ which follows from (iv), gives us $|P\rangle$ expressed in terms of the basic kets $|r\rangle$ with coefficients $\langle r|P\rangle$. These coefficients are thus the coordinates of $|P\rangle$. Similarly a bra $\langle Q|$ can be expressed in terms of the basic set $\langle s|$ with coefficients $\langle Q|s\rangle$ which are thus the coordinates of $\langle Q|$.

The scalar product

$$\langle Q|P\rangle = \sum_r \langle Q|r\rangle\langle r|P\rangle$$

in terms of the coordinates of $\langle Q|$ and $|P\rangle$. The coordinates of $|P\rangle$ are the conjugates of the coordinates of $\langle P|$.

A linear operator α has a double array of coordinates $\langle r|\alpha|s\rangle$ which can be conveniently written as a square matrix

$$\begin{pmatrix} \langle 1|\alpha|1\rangle & \langle 1|\alpha|2\rangle & \langle 1|\alpha|3\rangle & \cdots \\ \langle 2|\alpha|1\rangle & \langle 2|\alpha|2\rangle & \langle 2|\alpha|3\rangle & \cdots \\ & \vdots & \vdots & \\ \end{pmatrix}$$

Thus dynamical variables correspond to matrices. The condition to be satisfied so that α represent a real dynamical variable is $\alpha = \bar{\alpha}$ so that $\langle r|\alpha|s\rangle = \overline{\langle s|\alpha|r\rangle}$. Hence the matrix elements which are mirror images with respect to the principal diagonal are to be complex conjugates. Such matrices are called hermitean matrices. Thus real dynamical variables correspond to hermitean matrices.

From condition (iv) we can get $\langle r|\alpha\beta|s\rangle = \langle r|\alpha|t\rangle\langle t|\beta|s\rangle$ which is simply the law of matrix multiplication; hence the product of two dynamical variable corresponds to the product of the corresponding matrices. The coordinates of $\alpha|P\rangle$ are

$$\langle r|\alpha|P\rangle = \sum_s \langle r|\alpha|s\rangle\langle s|P\rangle$$

which is again the rule of matrix multiplication, provided we consider the coordinates of $|P\rangle$ as a single column matrix

$$\begin{pmatrix} \text{coordinates} \\ \text{of } \alpha \end{pmatrix} \begin{pmatrix} \text{coord} \\ \text{of} \\ |P\rangle \end{pmatrix} = \begin{pmatrix} \text{coord} \\ \text{of} \\ \alpha|P\rangle \end{pmatrix}$$

Similarly for a bra, the coordinates are to be considered as a single row matrix.

$$\langle Q|\alpha|r\rangle = \sum_s \langle Q|s\rangle\langle s|\alpha|r\rangle \,.$$

The linear operator $|A\rangle\langle B|$ has coordinates

$$\langle r|A\rangle\langle B|s\rangle$$

which is the matrix product of a single column matrix into a single row matrix. The product in the reverse order is just a number.

Consider the equation

$$\alpha|P\rangle = a|P\rangle ,$$

where α is a linear operator, a is a number and $|P\rangle$ is a non zero ket: $|P\rangle \neq 0$. In this equation, we consider α as a given linear operator and the ket $|P\rangle$ and the number a are unknown. We define a to be an eigenvalue of α and $|P\rangle$ to be an eigenket of α belonging to this eigenvalue. The property of being an eigenket depends only on the direction of the ket; any non-zero multiple of an eigenket is again an eigenket belonging to the same eigenvalue. If we have two eigenkets belonging to the same eigenvalue, any linear combination of them also will be an eigenket belonging to the same eigenvalue.

Since the formulation is symmetric with respect to bras and kets, we consider also the eigenvalue equation

$$\langle Q|\alpha = b\langle Q|$$

where b is a number and $\langle Q| \neq 0$. Thus we have, for every linear operator α , two types of eigenvalues associated with kets and bras respectively.

Assume α to be real: $\tilde{\alpha} = \alpha$ and assume $\alpha|P\rangle = a|P\rangle$. Multiplying both sides by $\langle P|$

$$\langle P|\alpha|P\rangle = a\langle P|P\rangle$$

Now $\langle P|P\rangle$ is real and not equal to zero. $\langle P|\alpha|P\rangle$ is also real. Hence a is real. Thus the eigenvalues of a real linear operator are real numbers.

The conjugate equation is

$$\langle P|\alpha = a\langle P|$$

Hence the eigenvalues are the same for both types and eigenbras are the conjugates of the eigenkets.

A number k is a linear operator. Take k in place of α . In this case any ket is an eigenket and there is only one eigenvalue namely k. Conversely, if we have a linear operator with only one eigenvalue and every ket is an eigenket, then it is simply a multiplicative constant.

Orthogonality Theorem: If we have two eigenvalues a and b with eigenkets $|A\rangle$ and $|B\rangle$

$$\alpha|A\rangle = a|A\rangle ,$$

$$\alpha|B\rangle = b|B\rangle .$$

We take conjugate of the second equation to obtain

$$\langle B|\alpha = b\langle B|$$

and we get by multiplication

$$\langle B|\alpha|A\rangle = a\langle B|A\rangle = b\langle B|A\rangle$$

Hence if $a \neq b$, $\langle B|A \rangle = 0$. This orthogonality relation is a very useful result.

Lecture 3

The question of whether a given linear operator has eigenvalues and eigenvectors, and, if so, of finding them is in general a difficult one. But for the special case when the linear operator α, say, satisfies an algebraic equation, e.g.,

$$F(\alpha) \equiv c_0\alpha^n + c_1\alpha^{n-1} + \dots + c_n = 0, \qquad (1)$$

one can easily obtain the eigenvalues and eigen-kets. If $|P\rangle$ is an eigen-ket of α belonging to the eigenvalue a then for any power series $f(\alpha)$ of α one has

$$f(\alpha)\,|P\rangle = f(a)\,|P\rangle .$$

In particular one would have

$$F(\alpha)\,|P\rangle = F(a)\,|P\rangle .$$

So from (1)

$$F(a) = 0 .$$

Thus the eigenvalues of α are roots of equation (1).

One can easily establish the following results.

(A) if $F(\alpha) = 0$ is the simplest equation satisfied by α then every root of $F(a) = 0$ is an eigenvalue of α.

(B) There exist so many eigenkets of α that they span the whole space.

For example if the operator α satisfies the equation

$$\alpha^2 = 1$$

then $(1+\alpha)|P\rangle$ and $(1-\alpha)|P\rangle$, $|P\rangle$ being any ket, will be eigenkets of α belonging to the eigenvalues 1 and − 1.

respectively provided they do not vanish. The arbitrary ket $|P\rangle$ can be written as a linear combination of these eigenkets of α :

$$|P\rangle = \frac{1}{2}(1+\alpha)|P\rangle + \frac{1}{2}(1-\alpha)\,|P\rangle .$$

The case of general linear operators has been treated by J. von Neumann in a lengthy article in Mathematische Annalen, 52, (1929), p 99 and by M.H. Stone in the book Linear Transformations in Hilbert Space. They have shown that for certain types of operators, called self adjoint operators, there exist so many eigenvectors that they span the whole space. The set of eigenvalues, which is called the spectrum, can comprise discrete numbers or continuous values in a (finite or infinite) interval, or both.

An eigenket of an operator belonging to an eigenvalue* α' in a range, (a, b), say, must be of infinite length, as the following argument shows. One forms the ket

$$|X\rangle = \int_a^b X_{\alpha'}\,|\alpha'\rangle\, d\alpha' ,$$

where $X_{\alpha'}$ is some function of α'. This is a ket of a general kind and the scaler product

$$\langle\alpha''|X\rangle = \int_a^b X_{\alpha'}\,\langle\alpha''|\alpha'\rangle\, d\alpha'$$

will therefore not be zero except in special cases. Now because of the orthogonality theorem the integrand is zero except when $\alpha' = \alpha''$, and

* Primed greek letters will always be taken to denote the eigenvalues of the operators which are themselves designated by the corresponding unprimed letters.

if $\langle\alpha'|\alpha''\rangle$ were finite the integral would be zero. One must therefore regard $\langle\alpha'|\alpha''\rangle$ to have a singularity at $\alpha' = \alpha''$ such that the integral is non-zero.

To represent this singularity one introduces a function defined by

$$\left.\begin{aligned} \delta(x) &= 0 \quad \text{for } x \neq 0, \\ \int \delta(x)\, dx &= 1, \end{aligned}\right\} \tag{2}$$

the range of integration being any interval containing the origin. The singular function $\langle\alpha''|\alpha'\rangle$ can now be represented as

$$\langle\alpha''|\alpha'\rangle = k(\alpha')\,\delta(\alpha'-\alpha'') \tag{3}$$

The function $\delta(x)$ defined above does not exist in a rigorous mathematical sense, and has been criticised by mathematicians on that account. One could do without it if one used the methods of von Neumann and Stone. But it constitutes a very useful notation and in physics one has now got used to it.

One can easily verify the following properties of the δ-function:

$$\int f(x)\,\delta(x)\,dx = f(0), \tag{4}$$

$$x\,\delta(x) = 0. \tag{5}$$

If one divides the equation $A = B$ by a continuous variable x that can take the value zero, one gets

$$\frac{A}{x} = \frac{B}{x} + k\,\delta(x) \tag{6}$$

in general, and not $A/x = B/x$.

Let us re-examine the proof of the orthogonality theorem for the case when the eigenvalues lie in a continuous range. If $|A\rangle$ and $|B\rangle$ are eigenvectors of a real operator belonging respectively to eigenvalues a and b in a range, then

$$\langle B|\alpha|A\rangle = a\langle B|A\rangle = b\langle B|A\rangle,$$

so that

$$(b - a) \langle B|A \rangle = 0,$$

whence on dividing by (b - a) one gets, observing equation (6) above,

$$\langle B|A \rangle = k \, \delta(b - a) .$$

This is just equation (3) again.

To connect the mathematical theory developed here with physical measurements, it will now be assumed that if $|P\rangle$ is an eigenket of an operator α belonging to the eigenvalue a,

$$\alpha |P\rangle = a |P\rangle ,$$

then the measurement of the dynamical variable represented by α in the state $|P\rangle$ is certain to lead to the result a, and, conversely, if the measurement of α in a state $|P\rangle$ certainly gives the result a then $|P\rangle$ is an eigenvector of α belonging to the eigenvalue a. These assumptions refer to special conditions, for in general the measurement of dynamical variables for a given state yields various results with certain probabilities.

It is further reasonable to assume that all states which can be attained physically are of finite length and that states of infinite length cannot be attained. Thus, for example, a state representing a particle whose position is definitely fixed cannot be achieved for the particle could have infinite energy. Or again a particle with a precise value of momentum could be anywhere in the universe and would not correspond to a physically attainable state.

A question which can now be considered is whether every linear operator corresponds to something that can be measured. Firstly, the linear operator must be real, because the results of physical measurements are always real numbers. One cannot in general measure a complex dynamical variable by measuring its real and pure imaginary parts because this would require the measurement of two quantities, and the two measurements will usually interfere with one another.

When we measure a real linear operator we are at the same time measuring all functions of it. When α is found to have the value a, then $f(\alpha)$ must have the value $f(a)$. The function f here may be any function whose domain of definition includes all the eigenvalues of α. Thus it is necessary to assume that there exist general functions of the real operator if it is to correspond to something measurable. This imposes a condition on the real linear operator, because it is not true of the general linear operator. When a real linear operator satisfies this condition it is called an <u>observable</u>.

The following conditions are assumed to be satisfied by the general functions of an observable :

(i) If $\alpha|P\rangle = a|P\rangle$ then $f(\alpha)|P\rangle = f(a)|P\rangle$

(ii) Functions of α can be combined according to ordinary commutative algebra.

(iii) Anything that commutes with α also commutes with $f(\alpha)$.

The assumption that an eigenket $|A\rangle$ of an observable α belonging to the eigenvalue a represents a state in which the measurement of α is certain to lead to the result a provides a physical interpretation applicable in special cases. To get a general physical interpretation, it will now be assumed that :-

If a state P is represented by a normalized ket $|P\rangle$ then the average value of an observable α for the state P is $\langle P|\alpha|P\rangle$.

This is a real number since α is a real operator. This general physical interpretation is consistent with the former one, which can in fact be derived from it. It is a powerful assumption, for it applies also to functions of α and gives sufficient information for one to be able to calculate the probabilities of obtaining various results for a measurement of α for the given state P.

In general the measurement of two observables will interfere with each other: the results obtained by first measuring one and then immediately afterwards the other will in general be different from those obtained if the order of measuring the observables were reversed. In special cases, however, the measurement of one observable α, say, may not alter the frequency distribution for the measurement of the other (β) and conversely. The observables α and β are then said to be <u>compatible</u>. The measurement of $\alpha\beta$ would then give the same result as that of $\beta\alpha$ for all states:

$$\langle P|\,(\alpha\beta-\beta\alpha)\,|P\rangle = 0 \quad \text{for all } |P\rangle$$

so that

$$\alpha\beta = \beta\alpha .$$

Conversely it is easily seen that two commuting observables can be assumed to be compatible without inconsistency. The definition of compatibility can be directly extended to more than two observables, and the condition of compatability is always equivalent to the condition of commutation.

General functions of compatible observables can also be defined. They satisfy the following conditions :-

(i) if $\alpha, \beta, \ldots$ commute and if

$$\alpha\,|P\rangle = a\,|P\rangle, \qquad \beta\,|P\rangle = b\,|P\rangle, \;\ldots$$

then

$$f(\alpha,\beta,\ldots)\,|P\rangle = f(a,b,\ldots)\,|P\rangle .$$

(ii) functions of $\alpha, \beta, \cdots$ can be combined according to the laws of ordinary commutative algebra

(iii) Anything that commutes with $\alpha, \beta, \cdots$ also commutes with $f(\alpha, \beta, \cdots)$.

To fix a state in quantum mechanics one has to fix a direction in the space of ket rectors. For this purpose one takes a set of commuting observables $\xi_1, \xi_2, \ldots$ such that none is a function of the remaining ones, and which is the maximum set having this property, i.e., which is such that there does not exist an observable which commutes with all of them and is not a function of them. Such a set is called a <u>complete</u> set of commuting observables. A state is now fixed if it is a simultaneous eigen-state of a complete set of commuting observables. This contrasts with the situation in classical mechanics where one had to specify <u>all</u> the variables to fix a state.

Lecture 4

We now take up the question of representations. The chief new feature in this presentation of the basic theory is the method of introducing the representation.

Consider a set of commuting dynamical variables $\xi_1, \xi_2, \cdots$ and suppose we have a ket $|S\rangle$ such that every ket $|P\rangle$ can be represented as a function of the ξ 's times $|S\rangle$, say

$$|P\rangle = \psi(\xi)\,|S\rangle .$$

Then we may say that ket $|P\rangle$ is represented by $\psi(\xi)$. For the representation considered we prove the following two theorems:

Theorem 1. The representation is unique, i.e., there cannot be two different functions $\psi_1(\xi), \psi_2(\xi)$ such that

$$|P\rangle = \psi_1(\xi)\,|S\rangle = \psi_2(\xi)\,|S\rangle$$

For, if there are two such functions, by subtraction we obtain,

$$(\psi_1 - \psi_2)\,|S\rangle = 0$$

Take any ket $|Q\rangle$ and put

$$|Q\rangle = f(\xi)\,|S\rangle .$$

Then

$$\begin{aligned}(\psi_1 - \psi_2)\,|Q\rangle &= (\psi_1 - \psi_2)\,f\,|S\rangle \\ &= f\,(\psi_1 - \psi_2)\,|S\rangle = 0\end{aligned}$$

(Since the functions f and ψ commute). Hence $\psi_1 - \psi_2$ operating on any ket gives zero, so that $\psi_1 - \psi_2 = 0.$

Theorem 2. The set of commuting dynamical variables ξ is complete.

Take any dynamical variable ω which commutes with all the ξ 's.

Put

$$\omega|S\rangle = \alpha(\xi)\,|S\rangle .$$

For any $|P\rangle = \psi(\xi)\,|S\rangle$, we have

$$\begin{aligned}\omega|P\rangle &= \omega\,\psi(\xi)\,|S\rangle = \psi(\xi)\,\omega|S\rangle \\ &= \psi(\xi)\,\alpha(\xi)\,|S\rangle = \alpha(\xi)\,\psi(\xi)\,|S\rangle \\ &= \alpha(\xi)\,|P\rangle\end{aligned}$$

so that

$$\omega = \alpha(\xi) .$$

Hence any variable that commutes with all the ξ 's is a function of them and hence the set of commuting dynamical variables is complete.

So far we have not specified what type of functions are to be considered. The nature of the representation would depend on this choice. In practice two kinds of representations are important.

(i) Ordinary Representation.

The ξ 's are observables and $\psi(\xi)$ can be any functions defined over the domain of eigenvalues.

(ii) Fock Representation.

The ξ 's are not observables and in general are not even real. The functions $\psi(\xi)$ are restricted to be power series in positive powers.

The first mentioned representation is better suited for problems involving only a few particles (e.g. energy levels of an atom or molecule) while the Fock representation is preferable for many particle systems and fields.

The chosen ket $|S\rangle$ is called the standard ket of the representation. We can change the representation by changing the standard ket without changing the ξ's. Choose a new ket $|S^*\rangle$. Since any ket is expressible in terms of $|S^*\rangle$, in particular $|S\rangle$ is so expressible, say

$$|S\rangle = k(\xi)\,|S^*\rangle .$$

Any ket $|P\rangle$ will have different representers ψ, ψ^* such that

$$|P\rangle = \psi(\xi)\,|S\rangle \qquad \text{original representation}$$

$$|P\rangle = \psi^*(\xi)\,|S^*\rangle \qquad \text{new representation}$$

But since $|S\rangle$ is expressible in terms of $|S^*\rangle$

$$|P\rangle = \psi k\,|S^*\rangle$$

or

$$\psi^* = \psi k$$

Hence with the change of standard ket from $|S\rangle$ to $|S^*\rangle$, every representer gets multiplied by the same function k. This is a simple change; so the change in the standard ket does not give rise to any important change in the representation.

k so introduced should have a reciprocal k^{-1} since $|S^*\rangle$ must be expressible in terms of $|S\rangle$ in the form

$$|S^*\rangle = k^{-1}\,|S\rangle .$$

Hence K should not vanish anywhere in the domain of eigenvalues in the case of the ordinary representation. For the Fock representation the requirement is that there should exist a reciprocal power series for the power series corresponding to K.

$\psi(\xi)$ is call the wave function because in the early examples of quantum mechanics it did represent waves. However, in the general theory, there may be no connection between the function $\psi(\xi)$ and waves.

Ordinary Representation: Take any function $f(\xi)$ of the dynamical variables and form the quantity

$$\langle S | f(\xi) | S \rangle = F$$

which is a number depending linearly on $f(\xi)$. If $f(\xi)$ is positive definite, F is real and positive; for, suppose f is positive definite. Then for some suitable function $g(\xi)$ we can put

$$f(\xi) = \bar{g}(\xi)\, g(\xi).$$

Then

$$F = \langle S | \bar{g}(\xi)\, g(\xi) | S \rangle > 0$$

since it is theproduct of the ket $g(\xi)|S\rangle$ and its conjugate bra $\langle S| \bar{g}(\xi)$. F is hence of the type

$$F = \underset{\substack{\text{discrete} \\ \text{spectrum}}}{\sum} \rho_\xi f(\xi) + \underset{\substack{\text{Continuous} \\ \text{spectrum}}}{\int} \rho(\xi) f(\xi)\, d\xi$$

where ρ_ξ, $\rho(\xi)$ are positive numbers independent of f. ρ is called the weight-function of the representation.

When the standard ket is changed, the weight-function is also changed. By the definition of ρ

$$(1) \quad \langle S|f|S\rangle = \sum_\xi \rho_\xi f(\xi) + \int \rho(\xi) f(\xi)\, d\xi$$

and the new weight-function ρ^* satisfies by definition

$$(2) \quad \langle S^*|f|S^*\rangle = \sum_\xi \rho^*_\xi f(\xi) + \int \rho^*(\xi) f(\xi)\, d\xi$$

Since

$$|S\rangle = k(\xi)\, |S^*\rangle$$

we have

$$\langle S|f|S\rangle = \langle S^*|\bar{k} f k|S^*\rangle$$
$$= \sum_\xi \rho^*_\xi \bar{k} f k + \int \rho^*(\xi)\, \bar{k} f k\, d\xi$$

from (2) with $\bar{k} f k$ substituted for f. Comparing this with (1), we get

$$\rho = \rho^* \bar{k} k = \rho^* |k|^2$$

so that the change in the weight function is multiplication by a positive factor. Hence by choosing a suitable k, i.e. a suitable standard ket, we can make the weight function unity. This is usually done in practice. When this is done, the standard ket is fixed apart from a phase factor of the type

$$k = e^{i\gamma(\xi)}$$

where $\gamma(\xi)$ is some real function. We now have the wave functions fixed apart from a phase factor.

Hence for an ordinary representation, one takes a complete set of commuting observables ξ and chooses the standard ket so as to make the weight function unity. The standard ket and the wavefunctions are now fixed upto a phase factor.

Ordinary Representation with weight factor unity: When the weight factor has been reduced to unity by a suitable choice of the standard ket,

$$\langle S|f(\xi)|S\rangle = \sum_{\text{discrete}} f(\xi) + \int_{\text{continuous}} f(\xi)\, d\xi)$$

Consider two kets $|1\rangle$ and $|2\rangle$ with representers ψ_1 and ψ_2 :

$$|1\rangle = \psi_1(\xi)\ |S\rangle ,$$

$$|2\rangle = \psi_2(\xi)\ |S\rangle .$$

The scalar product is

$$\langle 1|2\rangle = \langle S|\bar{\psi}_1 \psi_2|S\rangle = \sum_{\text{discrete}} \bar{\psi}_1 \psi_2 + \int_{\text{Cont}} \bar{\psi}_1 \psi_2\, d\xi .$$

As an example, if the ket $|1\rangle$ is normalised,

$$\sum \bar{\psi}_1 \psi_1 + \int \bar{\psi}_1 \psi_1\, d\xi = 1$$

In this case we say that the wavefunction is normalised.

Suppose $|P\rangle$ is normalised and has the representative

$$|P\rangle = \psi(\xi)\ |S\rangle$$

Then the average of any function of the ξ 's is

$$\langle P|f|P\rangle = \langle S|\bar{\psi} f \psi|S\rangle = \sum f |\psi|^2 + \int f |\psi|^2 d\xi$$

From this expression for the average of $f(\xi)$, we see that the probability of the ξ 's having discrete values ξ' is $|\psi(\xi')|^2$ and the probability of their having values lying in the ranges ξ' to $\xi' + d\xi'$ is $|\psi(\xi')|^2 d\xi'$.

Hence provided the wavefunction is normalised, the square of the modulus of the wavefunction is the probability distribution function. This is the major advantage of the ordinary representation. No such direct physical interpretation exists for the Fock representation. In this case one has to transform back to an ordinary representation to get a physical interpretation.

ECG: ks
23.3.55

Lecture 5

The representation of kets introduced in the last lecture goes over in certain cases into the coordinates introduced earlier. Let us consider an ordinary representation generated by means of the observables $\xi_1, \xi_2, \ldots$, having each a discrete eigenvalue spectrum and the standard ket $|S\rangle$. Any ket $|P\rangle$ is then represented as

$$|P\rangle = \psi(\xi)\,|S\rangle\,. \tag{1}$$

The function $\psi(\xi)$ will in this case be a set of numbers, and may be written in the form of a single-column matrix. A dynamical variable acting on $|P\rangle$ would change it into another ket which could be similarly represented by another set of numbers arranged in a column. The dynamical variable is then represented as a square matrix.

In the last lecture we had for the scalar product of two kets $|1\rangle$ and $|2\rangle$, in a representation with weight factor unity,

$$\langle 1|2\rangle = \sum \bar{\psi}_1(\xi)\,\psi_2(\xi)\,. \tag{2}$$

The right hand side of (2) is the scalar product of the row representing $\langle 1|$ and the column representing $|2\rangle$ in a coordinate system. This shows that the multiplication law for the representatives is just matrix multiplication and hence the representatives are the same as the coordinates we had earlier.

We consider the example where there are only three operators satisfying the following rules:

$$\begin{gathered}\sigma_1\sigma_2 = -\sigma_2\sigma_1 = i\sigma_3\,; \quad \sigma_2\sigma_3 = -\sigma_3\sigma_2 = i\sigma_1\,, \\ \sigma_3\sigma_1 = -\sigma_1\sigma_3 = i\sigma_2\,, \\ \sigma_1^2 = \sigma_2^2 = \sigma_3^2 = 1\,. \end{gathered} \tag{3}$$

The eigenvalues of each one of these operators are ± 1. We consider the ordinary representation introduced by σ_3 3 and a standard ket $|S\rangle$ which we take to be an eigenket of σ_1 belonging to the eigenvalue $+1$,

$$\sigma_1 |S\rangle = |S\rangle \tag{4}$$

It is now easily verified that every ket $|P\rangle$ is expressible as

$$|P\rangle = \psi(\sigma_3)|S\rangle . \tag{5}$$

For example:

$$\left.\begin{aligned} \sigma_1 |S\rangle &= \psi(\sigma_3)|S\rangle = |S\rangle \quad \text{with } \psi(\sigma_3) = 1, \\ \sigma_3 |S\rangle &= \psi(\sigma_3)|S\rangle \quad \text{with } \psi(\sigma_3) = \sigma_3 , \\ \sigma_2 |S\rangle &= -i\,\sigma_3\sigma_1 |S\rangle = -i\sigma_3 |S\rangle = \psi(\sigma_3)|S\rangle \\ &\qquad \text{with } \psi(\sigma_3) = -i\sigma_3 \end{aligned}\right\} \tag{6}$$

and so on. Since the eigenvalues of the operator σ_3 are $+1$ and -1, the representative of every ket is merely a set of two numbers, which may be arranged in a column. We obtain:

$$|S\rangle = \begin{pmatrix} 1 \\ 1 \end{pmatrix}, \quad \sigma_3 |S\rangle = \begin{pmatrix} 1 \\ -1 \end{pmatrix}, \quad \sigma_2 |S\rangle = \begin{pmatrix} i \\ -i \end{pmatrix} . \tag{7}$$

With $|P\rangle = \begin{pmatrix} a \\ b \end{pmatrix}$ we have $\sigma_3 |P\rangle = \begin{pmatrix} a \\ -b \end{pmatrix}$ so we see that we can represent σ_3 as

$$\begin{pmatrix} 1 & 0 \\ 0 & -1 \end{pmatrix} . \tag{8a}$$

Let

$$\sigma_1 = \begin{pmatrix} p & q \\ r & s \end{pmatrix}.$$

Then from

$$\sigma_1 |S\rangle = |S\rangle$$

$$\sigma_1 \sigma_3 |S\rangle = -\sigma_3 \sigma_1 |S\rangle = -\sigma_3 |S\rangle$$

we get

$$\begin{pmatrix} p & q \\ r & s \end{pmatrix}\begin{pmatrix} 1 \\ 1 \end{pmatrix} = \begin{pmatrix} 1 \\ 1 \end{pmatrix}, \quad \begin{pmatrix} p & q \\ r & s \end{pmatrix}\begin{pmatrix} 1 \\ -1 \end{pmatrix} = \begin{pmatrix} -1 \\ 1 \end{pmatrix},$$

whence $p = 0,\ q = 1,\ r = 1,\ s = 0$, so that

$$\sigma_1 = \begin{pmatrix} 0 & 1 \\ 1 & 0 \end{pmatrix}. \tag{8b}$$

Similarly we obtain

$$\sigma_2 = \begin{pmatrix} 0 & -i \\ i & 0 \end{pmatrix} \tag{8c}$$

The matrices (8) are called the Pauli matrices.

We shall now introduce the operators of differentiations. Let us consider an ordinary representation generated by dynamical variables $q_1, q_2, \ldots,$ having eigenvalues in a continuous range, and the standard ket $|S\rangle$. We have, for any ket $|P\rangle$

$$|P\rangle = \psi(q)\ |S\rangle. \tag{9}$$

Then $\frac{\partial \psi}{\partial q_r} |S\rangle$ is a ket depending linearly on $\psi|S\rangle$, and may be considered as the result of the linear operator ∂_r operating on $\psi|S\rangle$:

$$\frac{\partial \psi}{\partial q_r} |S\rangle = \partial_r\ \psi|S\rangle. \tag{10}$$

We must see whether we can give a meaning to ∂_r operating to the left on a bra $\langle Q|$, i.e. whether we can give a meaning to $\langle Q|\partial_r$. Put

$$\langle Q| = \langle S|\varphi(q). \tag{11}$$

Then we require that

$$(\langle S|\varphi\partial_r)\psi|S\rangle = \langle S|\varphi(\partial_r\psi|S\rangle). \tag{12}$$

If the representation has weight factor unity, then the right hand side of the above equation is

$$\begin{aligned}\langle S|\varphi(\partial_r\psi|S\rangle) &= \langle S|\varphi\frac{\partial\psi}{\partial q_r}|S\rangle \\ &= \int\varphi\frac{\partial\psi}{\partial q_r}\,dq.\end{aligned} \tag{13}$$

Let

$$\langle S|\varphi\partial_r = \langle S|\chi. \tag{14}$$

Then the left hand side of (12) is

$$(\langle S|\varphi\partial_r)\psi|S\rangle = \langle S|\chi\psi|S\rangle = \int\chi\psi\,dq. \tag{15}$$

We thus require a function χ such that

$$\int\chi\psi\,dq = \int\varphi\frac{\partial\psi}{\partial q_r}\,dq. \tag{16}$$

Such a function does exist if the wave function ψ vanishes at the boundaries, and is then

$$\chi = -\frac{\partial\varphi}{\partial q_r} \tag{17}$$

Hence the operator ∂_r acting on bra vectors has the meaning of minus the differentiation operator if suitable boundary conditions are imposed on the wave functions,

$$\langle S|\varphi\,\partial_r = -\langle S|\,\frac{\partial\varphi}{\partial q_r}.$$

The conjugate imaginary of $\partial_r\,\psi|S\rangle$ is $\langle S|\,\bar{\psi}\,\bar{\partial}_r$. But $\partial_r\,\psi|S\rangle = \frac{\partial\psi}{\partial q_r}|S\rangle$ and its conjugate imaginary is $\langle S|\,\frac{\partial\bar{\psi}}{\partial q_r}$, so that

$$\langle S|\,\bar{\psi}\,\bar{\partial}_r = \langle S|\,\frac{\partial\bar{\psi}}{\partial q_r} = -\langle S|\,\bar{\psi}\,\partial_r$$

Hence $\bar{\partial}_r = -\partial_r$, and ∂_r is pure imaginary.

We now wish to find the commutation relation between ∂_r and $f(q)$

We have

$$\partial_r\,f(q)\,|P\rangle = \partial_r\,f\,\psi|S\rangle$$

$$= \frac{\partial}{\partial q_r}(f\psi)\,|S\rangle = \frac{\partial f}{\partial q_r}\psi|S\rangle + f\,\frac{\partial\psi}{\partial q_r}|S\rangle$$

$$= \frac{\partial f}{\partial q_r}|P\rangle + f\,\partial_r\,|P\rangle,$$

whence $\quad \partial_r f - f\,\partial_r = \frac{\partial f}{\partial q_r},$

and, in particular,

$$\partial_r\,q_s - q_s\,\partial_r = \delta_{rs}. \tag{18}$$

Also, as is directly seen,

$$\partial_r\,\partial_s = \partial_s\,\partial_r. \tag{19}$$

The relation

$$\langle S|\varphi\,\partial_r = -\langle S|\,\frac{\partial\varphi}{\partial q_r}$$

is correct only when the wave function of the ket vector with which this bra is multiplied satisfies the boundary conditions. Consider, for example,

$$\langle S| \partial_r q_s - q_s \partial_r |S\rangle = \delta_{rs} \langle S|S\rangle .$$

Now $\partial_r |S\rangle = 0$, and its conjugate imaginary is $-\langle S| \partial_r = 0$ so that the left hand side of the above equation is zero. This would lead to

$$\langle S|S\rangle = 0$$

which is impossible. This inconsistency is due to the fact that $\langle S| \partial_r$ in the above equation is multiplied with $q_s |S\rangle$, whose wave function q_s des <u>not</u> satisfy the boundary condition.

We shall now introduce the quantum conditions and show that the linear operators ∂_r have a physical meaning. Consider a dynamical system consisting of a system of particles, described classically by the coordinates $q_1, q_2, \ldots$, and the conjugate momenta $p_1, p_2, \ldots$. In classical mechanics all these dynamical variables are assumed to commute with one another. In quantum mechanics we need to replace this assumption by some other rules since the dynamical variables are now linear operators. We postulate that they satisfy the following <u>quantum conditions</u>:

$$\left.\begin{aligned} q_r q_s - q_s q_r &= 0 , \\ p_r p_s - p_s p_r &= 0 , \\ q_r p_s - p_s q_r &= i\hbar \delta_{rs} \end{aligned}\right\} \qquad (20)$$

where $\hbar = \frac{h}{2\pi}$, h being Planck's constant. In the last equation i is needed on the right hand side because $q_r p_s - p_s q_r$ is a pure imaginary operator.

The q's form a set of commuting observables, which is evidently complete, so we can set up a representation with the wave functions of the q's. In this representation we can introduce the operators of differentiation ∂_r .

We shall now show that

$$p_r = -i\,\partial_r \tag{21}$$

if it is assumed that the p_r's possesses the eigenvalues zero and if we take the standard ket to be an eigenket of all the p_r's belonging to the eigenvalues zero, so that

$$p_r\,|S\rangle = 0 .$$

Comparing

$$q_r p_s - p_s q_r = i\hbar\,\delta_{rs}$$

and

$$q_r \partial_s - \partial_s q_r = -\delta_{rs}$$

we see that

$$q_r (p_s + i\hbar\,\partial_s) - (p_s + i\hbar\,\partial_s)\, q_r = 0$$

so that

$$(p_s + i\hbar\,\partial_s) = f_s(q);$$

since $p_s\,|S\rangle = 0, \quad \partial_s\,|S\rangle = 0$, we have

$$f_s\,|S\rangle = 0$$

whence $f_s = 0$, which proves (21).

This representation is called the Schrödinger representation.

Lecture 6

The quantum conditions, introduced previously, are symmetric with respect to the interchange of coordinates and momenta, provided simultaneously i is replaced by -i. From this follows the possibility of representing the wave-functions as functions of momenta with the coordinates appearing as differential operators. If $|S_q\rangle$ and $|S_p\rangle$ denote the standard kets of the Schrödinger and momentum representations respectively any general ket $|P\rangle$ can be expressed in the forms:

$$|P\rangle = \psi(q)\,|S_q\rangle, \quad p_r\,|S_q\rangle = 0, \qquad \text{Schrödinger representation}$$

$$|P\rangle = \phi(p)\,|S_p\rangle, \quad q_r\,|S_p\rangle = 0. \qquad \text{Momentum representation}$$

Hence we must have the relation

$$\psi(q)\,|S_q\rangle = \varphi(p)\,|S_p\rangle$$

where $\varphi(p)$ depends linearly on $\psi(q)$.

Take the case of a single degree of freedom, so that there is only one p and one q. Let us suppose that $|p'\rangle$ is an eigenket of the momentum belonging to the eigenvalue p':

$$p\,|p'\rangle = p'\,|p'\rangle$$

Expressing this relation in terms of $|S_q\rangle$, with $|p\rangle = \psi(q)\,|S_q\rangle$

$$p\,\psi(q)\,|S_q\rangle = p'\,\psi(q)\,|S_q\rangle$$

In the Schrödinger representation, p is an operator of differentiation, $p = -i\hbar\dfrac{\partial}{\partial q}$, so that

$$-i\hbar\,\frac{\partial\psi}{\partial q}\,|S_q\rangle = p'\,\psi(q)\,|S_q\rangle.$$

Since the representative of a ket is unique, this demands that

$$-i\hbar \frac{\partial \psi}{\partial q} = p'\,\psi(q)$$

or

$$\psi(q) = a_{p'}\, e^{i q p'/\hbar}$$

where $a_{p'}$ is some function of p', independent of q (left undefined for the moment). $\psi(q)$ so obtained is the representative of the eigenket $|p'\rangle$ of momentum (belonging to the eigenvalue p') in the Schrödinger representation.

Similarly, if we express the eigenvalue equation for the ket in the momentum representation, we get

$$p\,\varphi(p)\,|S_p\rangle = p'\,\varphi(p)\,|S_p\rangle,$$

and since the representation is unique,

$$p\,\varphi(p) = p'\,\varphi(p)$$

which requires that

$$\phi(p) = \delta(p-p')$$

if we choose the arbitrary coefficient of proportionality on the right hand side to be unity by defining the eigenket $|p'\rangle$ suitably. The above gives the momentum representative of the eigenket $|p'\rangle$.

We thus see that a momentum representative $\varphi(p) = \delta(p-p')$ corresponds to a Schrödinger representative $\psi(q) = a_{p'}\, e^{iqp'/\hbar}$. A general momentum representative $\phi(p)$ can be written as an integral

$$\phi(p) = \int \phi(p')\,\delta(p-p')\,dp'$$

and would correspond to a Schrödinger representative

$$\psi(q) = \int \phi(p')\, a_{p'}\, e^{i q p'/\hbar}\, dp'$$

The Schrödinger representative is hence defined except for the unknown coefficent $a_{p'}$ surviving in the calculation.

The quantum conditions are symmetric with respect to interchange of q and p if -i is written for i everywhere. Adopting this interchange, we see that $\phi(p)$ should be expressible in terms of $\psi(q)$ in the form

$$\phi(p) = \int \psi(q')\, b_{q'}\, e^{-iq'p/\hbar}\, dq'$$

where $b_{q'}$ is an undefined function of q'. Taking the Fourier-inverses of the expressions for $\psi(q)$ and $\phi(p)$, we get

$$\psi(q)\cdot b_q = \frac{1}{h}\int \varphi(p)\, e^{iqp/\hbar}\, dp,$$

$$\phi(p)\, a_p = \frac{1}{h}\int \psi(q)\, e^{-iqp/\hbar}\, dq.$$

For the two linear connections written down between ψ and ϕ to be consistent, it is necessary to have $a_p = b_q = \frac{1}{h^{1/2}}$. The relation between $\psi(q)$ and $\phi(p)$ can now be written as

$$\psi(q) = \frac{1}{h^{1/2}}\int \varphi(p)\, e^{iqp/\hbar}\, dp,$$

$$\phi(p) = \frac{1}{h^{1/2}}\int \psi(q)\, e^{-iqp/\hbar}\, dq.$$

In these expressions, p and q are ordinary algebraic entities which commute (and not linear operators).

For the case of n degrees of freedom, the generalisation is immediate. $\frac{iqp}{\hbar}$ occurring in the exponents of the expressions relating ϕ and ψ are to be replaced by $i\sum \frac{q_r p_r}{\hbar}$, the simmuation to be extended over the various degrees of freedom, and the coefficient $h^{1/2}$ is to be replaced by $h^{-n/2}$.

Physically the Schrödinger and momentum wave functions have the interpretation that $|\psi(q)|^2$ gives the probability distribution for the coordinate and $|\varphi(p)|^2$ that for the momentum.

The formulae directly lead to Heisenberg's Principle of Uncertainty. If Δq measures the precision in the determination of the position, $\psi(q)$ may be represented by a wavepacket consisting of a train of approximately monochromatic waves extending for a length Δq, similar to the figure 1a, where the real part of $\psi(q)$ is depicted. The Fourier transform of such a $\psi(q)$ is a function $\varphi(p)$ which is also of the form of a wave packet. Compare figure 1b where the real part of $\varphi(p)$ is shown. If Δp is the range of momenta involved, then in order that the Fourier components of $\psi(q)$ at the two ends of the interval Δp may reinforce each other in the middle of the domain Δq and interfere with one another at the ends of the domain Δq, we must have

$$\Delta q \, \Delta p = h,$$

which is Heisenberg's Principle.

To discuss the relative merits of the Schrödinger and momentum representations, we note that in the Schrödinger representation q_r is a multiplying factor and p_r is a differential operator $-i\hbar \frac{\partial}{\partial q_r}$; while in the momentum representation p_r is a multiplying factor and q_r is a differential operator $i\hbar \frac{\partial}{\partial p_r}$. Hence in the Schrödinger representation any function of the dynamical variables which can be expressed as a power series in p is readily expressed. Similarly any function which is a power series in the coordinates is readily expressed in the momentum representation. The usual dynamical quantities like energy(*) are power series in the momentum variables. But for problems involving potentials (say the potential of the hydrogen atom $\sim \frac{1}{r}$

$$= \frac{1}{\sqrt{(q_1^2+q_2^2+q_3^2)}}\ )$$

we have functions of the coordinates which cannot be expressed as power series. In these cases the problems are best formulated in the Schrödinger representation. On the other hand, the momentum representation is usually more useful when physical interpretation of the theory is concerned because experimenters usually measure momenta and not coordinates. One can with advantage use both, formulating the problem in the Schrödinger representation and later transforming to the momentum representation.

Since the elementary examples are worked out in detail in various text books, we will not, in this lecture series, discuss them at length. However a particularly important case, that of the simple harmonic oscillator, being the brick out of which field theories are built, will be handled below by a method which is specially adapted for application to field theories.

Harmonic Oscillator with a single degree of freedom: Leaving out certain coefficients, the energy of a simple harmonic oscillator is

$$\frac{1}{2}(p^2+q^2)$$

(*) It is true that the relativistic Hamiltonian contains a square root of a function of p. However, the square root is resolved into a linear expression when spin variables are introduced.

In the Schrödinger representation

$$\frac{1}{2}(q^2+p^2) = \frac{1}{2}\left(q^2-\hbar^2\frac{\partial^2}{\partial q^2}\right)$$

and the eigenvalues satisfy the equation

$$\frac{1}{2}\left(q^2-\hbar^2\frac{\partial^2}{\partial q^2}\right)\psi(q) = \lambda\psi(q).$$

Instead of solving this equation, we shall assume a particular eigenvalue λ to exist and deduce other from it.

Let $|\lambda\rangle$ be the corresponding eigen ket. Then

$$\frac{1}{2}(q^2+p^2)|\lambda\rangle = \lambda|\lambda\rangle.$$

Now,

$$(q+ip)(q-ip) = q^2+p^2+i(pq-qp)$$
$$= q^2+p^2+\hbar$$

by virtue of the quantum conditions.

Similarly,

$$(q-ip)(q+ip) = q^2+p^2-\hbar.$$

Hence

$$\frac{1}{2}(q+ip)(q-ip)|\lambda\rangle = (\lambda+\tfrac{1}{2}\hbar)|\lambda\rangle$$

Multiplying both sides by $(q-ip)$ on the left

$$\frac{1}{2}(q-ip)(q+ip)(q-ip)|\lambda\rangle = (\lambda+\tfrac{1}{2}\hbar)(q-ip)|\lambda\rangle$$

Replacing the first two factors by $q^2+p^2-\hbar$ and transposing $-\frac{1}{2}\hbar(q-ip)|\lambda\rangle$ to the right, one obtains,

$$\frac{1}{2}(q^2+p^2)(q-ip)|\lambda\rangle = (\lambda+\hbar)(q-ip)|\lambda\rangle.$$

Hence $(q-ip)|\lambda\rangle$ is also an eigenket of $\frac{1}{2}(q^2+p^2)$ but belonging to the eigenvalue $\lambda+\hbar$ provided

$$(q-ip)|\lambda\rangle \neq 0.$$

By similar steps we can show that the ket $(q+ip)|\lambda\rangle$ is an eigenket of $\frac{1}{2}(q^2+p^2)$ belonging to the eigenvalue $\lambda-\hbar$

provided,

$$(q + ip)\,|\lambda\rangle \neq 0.$$

Hence starting with any one value λ , we can get an arithmetical progression of eigenvalues extending to either side and differing in steps of $\hbar$, provided at no step the two conditions stipulated above are violated.

The eigenvalues are to be all non-negative (i.e. zero or positive). To see this we note that the average value of q^2 for any state $|P\rangle$ is

$$\langle P|q^2|P\rangle = \langle P|q \cdot q|P\rangle \geqslant 0$$

and similarly the average value of p^2 also is non-negative; hence the average value of q^2+p^2 is non-negative. For an eigenstate, the average value is the eigenvalue and hence it follows that all eigenvalues are non-negative. Hence a minimum (non-negative) eigenvalue λ_{min} exists with its corresponding ket satisfying the relation

$$(q + ip)\;|\lambda_{min}\rangle = 0.$$

Multiplying on the left by $q - ip$ we obtain

$$(q - ip)\,(q + ip)\,|\lambda_{min}\rangle = 0.$$

But $(q - ip)(q + ip) = q^2 + p^2 - \hbar$ so that,

$$(\lambda_{min} - \tfrac{1}{2}\hbar)\;|\lambda_{min}\rangle = 0$$

giving

$$\lambda_{min} = \tfrac{1}{2}\hbar .$$

If there were a maximum eigen-value λ_{max}, the corresponding eigenket would satisfy the relation,

$$(q - ip)\,|\lambda_{max}\rangle = 0$$

Multiplying from the left by $(q + ip)$ and using similar steps we get

$$\lambda_{max} = -\frac{1}{2}\hbar$$

which violates the restriction demanding the eigenvalues to be all positive. Hence we conclude that no maximum eigenvalue exists and the eigenvalues form an arithmetical progression starting from $\frac{1}{2}\hbar$ and extending to infinity in steps of $\hbar$.

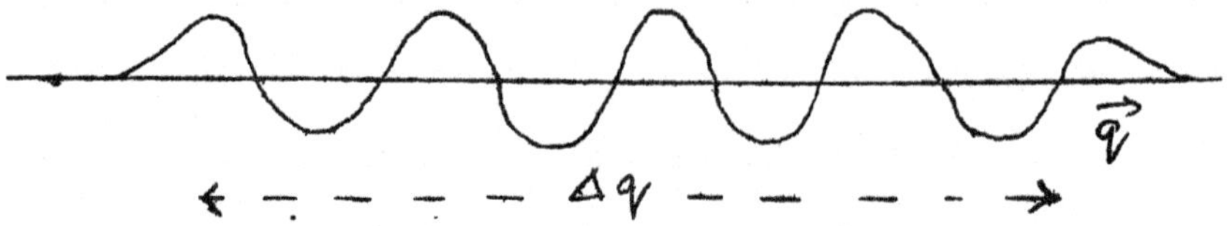

fig 1a real part of $\psi(q)$

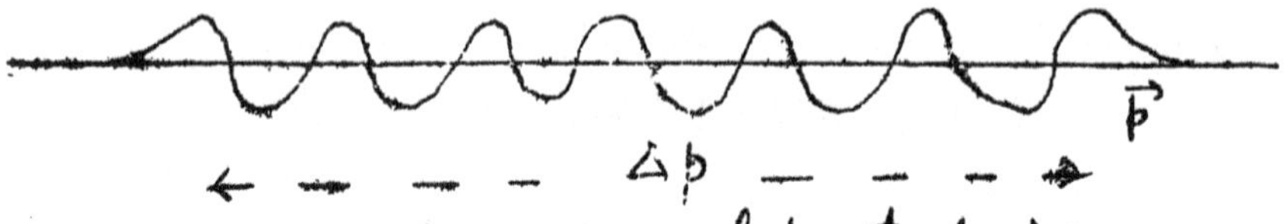

fig. 1b real part of $\phi(p)$

Lecture 7

We shall now consider a representation of the Fock type for dealing with the one dimensional harmonic oscillator. We introduce the complex dynamical variables

$$\eta = \frac{1}{\sqrt{2\hbar}}(q - ip) \tag{1}$$

and

$$\bar{\eta} = \frac{1}{\sqrt{2\hbar}}(q + ip). \tag{2}$$

We have

$$\begin{aligned}\bar{\eta}\eta - \eta\bar{\eta} &= \frac{1}{2\hbar}\{(q+ip)(q-ip) - (q-ip)(q+ip)\} \\ &= -\frac{1}{\hbar}\, i(qp - pq) = 1.\end{aligned} \tag{3}$$

Since there is nothing independent of η that commutes with η, it will by itself form a complete set, and we can set up a representation in which the wave functions are functions of η. As standard ket we take the eigenket $|S\rangle$ of H belonging to the lowest eigenvalue $\frac{1}{2}\hbar$. Then, as was shown in the last lecture,

$$\bar{\eta}\,|S\rangle = 0 \tag{4}$$

and

$$|S\rangle,\ \eta|S\rangle,\ \ldots\ \eta^n|S\rangle,\ \ldots. \tag{5}$$

belong respectively to the eigenvalues $1/2\,\hbar$, $3/2\,\hbar$, $5/2\,\hbar$, $\ldots$, $(n+\frac{1}{2})\hbar$, $\ldots$ respectively, of H. The kets (5) span the whole space since nothing new is obtained by multiplying $|S\rangle$ by p and q. A general ket will then be expressible as a linear combination of kets (5), i.e.,

$$|P\rangle = (a_0 + a_1\eta + a_2\eta^2 + \cdots)\,|S\rangle \equiv \psi(\eta)\,|S\rangle. \tag{6}$$

The wave function $\psi(\eta)$ representing $|P\rangle$ is thus a power series in η. This representation is very convenient when one is interested in the eigenstates of H, since the various terms in (6) are eigenfunctions of H.

In this Fock representation all linear operators are expressible in terms of η and $\bar{\eta}$. The operator η acts as a mere multiplying factor. To see the effect of $\bar{\eta}$ we note that

$$\bar{\eta}\eta^n - \eta^n\bar{\eta} = n\,\eta^{n-1}; \tag{7}$$

this relation is readily proved by induction from (3). Hence

$$\begin{aligned}\bar{\eta}|P\rangle &= \bar{\eta}\,\psi(\eta)\,|S\rangle \\ &= \{\bar{\eta}\,\psi(\eta) - \psi(\eta)\bar{\eta}\}\,|S\rangle \\ &= \sum_n a_n\, n\,\eta^{n-1}\,|S\rangle \\ &= \frac{\partial\psi}{\partial\eta}\,|S\rangle .\end{aligned} \tag{8}$$

In the second step we used the relation $\bar{\eta}|S\rangle = 0$. Thus $\bar{\eta}$ acting on a wave function just differentiates it, as indeed one would expect from the commutation relation (3).

The important difference between the Schrödinger representation and the Fock representation is that the dynamical variables $\eta, \bar{\eta}$, which appear as multiplying factor and operator of differentiation respectively in the latter representation, are complex, whereas the corresponding variables q, p in the Schrödinger representation are real. Moreover $|\psi(q)|^2$ in the Schrödinger representation has the physical meaning of being the probability distribution of q. The quantity $|\psi(\eta)|^2$ has no such interpretation in the Fock representation for η is not an observable at all.

An interpretation for $\psi(\eta)$ can however be obtained from its power series expansion. Assume that $|S\rangle$ is of length unity, $\langle S|S\rangle = 1$. Then for the squared length of $\eta^n|S\rangle$ we obtain

$$\langle S|\bar{\eta}^n \eta^n|S\rangle = \langle S|\bar{\eta}^{n-1}\bar{\eta}\,\eta^n|S\rangle$$
$$= n\langle S|\bar{\eta}^{n-1}\eta^{n-1}|S\rangle$$

since $\bar{\eta}$ is the operator of differentiation as was shown above. Repeating this process n times we obtain for the squared length of $\eta^n|S\rangle$

$$\langle S|\bar{\eta}^n \eta^n|S\rangle = n!\,\langle S|S\rangle = n!\,. \tag{9}$$

Also any two of the kets in (5) are orthogonal from the orthogonality theorem. We now obtain for the scalar product of two kets

$$|P\rangle = \psi(\eta)|S\rangle = \sum a_n \eta^n|S\rangle$$

and

$$|Q\rangle = \varphi(\eta)|S\rangle = \sum b_n \eta^n|S\rangle$$

the following expression:

$$\langle Q|P\rangle = \sum_{n,m} a_n \bar{b}_m \langle S|\bar{\eta}^m \eta^n|S\rangle$$
$$= \sum_n a_n \bar{b}_n \langle S|\bar{\eta}^n \eta^n|S\rangle$$
$$= \sum_n a_n \bar{b}_n\, n!\,. \tag{10}$$

If the ket $|P\rangle$ is normalized, then

$$\langle P|P\rangle = \sum_n |a_n|^2 n! = 1, \tag{11}$$

so that a_n must tend to zero quite fast as $n \to \infty$. $|a_n|^2 n!$ is now the probability that the state represented by $|P\rangle$ is the nth excited state of the harmonic oscillator, i.e., that H has the eigenvalue $(n+\frac{1}{2})\hbar$. This is the physical interpretation which one has for the Fock representation.

The above treatment of a single harmonic oscillator can be directly extended to the case of a number of them. One then gets a scheme which can be used for dealing with fields.

Consider a system constituted of a number of oscillators, the rth of them being described by the variables q_r, p_r, and having the energy

$$H_r = \frac{1}{2}\omega_r(q_r^2 + p_r^2) \tag{12}$$

The energy of the whole system is

$$H_T = \sum_r H_r . \tag{13}$$

It is evident from the quantum conditions that different degrees of freedom do not interfere, because dynamical variables referring to different degrees of freedom commute.

The eigenvalues of H_r will be

$$\frac{1}{2}\hbar\omega_r,\ \frac{3}{2}\hbar\omega_r, \ldots, (n+\frac{1}{2})\hbar\omega_r, \ldots . \tag{14}$$

We define

$$\left.\begin{aligned} \eta_r &= \frac{1}{\sqrt{2\hbar}}(q_r - ip_r), \\ \bar{\eta}_r &= \frac{1}{\sqrt{2\hbar}}(q_r + ip_r). \end{aligned}\right\} \tag{15}$$

Then

$$\bar{\eta}_r \eta_s - \eta_s \bar{\eta}_r = \delta_{rs} \tag{16}$$

so $\bar{\eta}_r$ is the operator of differentiation

$$\bar{\eta}_r = \frac{\partial}{\partial \eta_r} .$$

Let $|S\rangle$ be an eigenket of H_r (all r) satisfying

$$\bar{\eta}_r |S\rangle = 0 \qquad \text{for all } r.$$

Then $|S\rangle$ will be the eigenket of H_r belonging to the lowest eigenvalue $\frac{1}{2}\hbar\omega_r$ for all r. The ket

$$\eta_1^{n_1} \eta_2^{n_2} \ldots |S\rangle \tag{17}$$

will represent a state in which the ith oscillator is excited to n_i quanta, i.e., in which H_i has the eigenvalue $(n_i + \frac{1}{2})\hbar\omega_i$ for all i. A general ket $|P\rangle$ will be a linear combination of kets of form (17), so that

$$|P\rangle = \sum_{n_1, n_2, \ldots} a_{n_1, n_2, \ldots} \eta_1^{n_1} \eta_2^{n_2} \ldots |S\rangle$$

$$\equiv \psi(\eta) |S\rangle \tag{18}$$

with $\psi(\eta)$ a power series in all the η variables. If $|S\rangle$ is normalized, then the squared length of the ket (17) will be

$$n_1!\; n_2! \cdots .$$

This result follows directly from the corresponding one obtained for the case of a single oscillator if it is remembered that different degrees of freedom do not interfere. If the ket (18) is normalized then

$$n_1!\; n_2! \cdots |a_{n_1 n_2 \ldots}|^2$$

can be interpreted as the probability for the rth oscillator to be in the $n^{r}\,th$ excited state for all r .

If the number of oscillators be infinite we obtain a mathematical description of a quantized field. The Fock representation used here is the most convenient one for dealing with the problem.

We shall now consider some general features of fundamental importance concerning angular momentum. To define angular momentum we consider first a single particle described by the coordinates x_1 , x_2 , x_3 and the conjugate momenta p_1 , p_2 , p_3 . The angular momentum $\vec{m}$ of the particle is defined as

$$\left.\begin{aligned} m_1 &= x_2 p_3 - x_3 p_2 , \\ m_2 &= x_3 p_1 - x_1 p_3 , \\ m_3 &= x_1 p_2 - x_2 p_1 . \end{aligned}\right\} \tag{19}$$

To obtain the eigenvalues of m_1 we introduce the Schrödinger representation. Then

$$m_1 |P\rangle = -i\hbar\left(x_2 \frac{\partial}{\partial x_3} - x_3 \frac{\partial}{\partial x_2}\right) \psi(x_1 x_2 x_3)\, |S\rangle . \tag{20}$$

Now it is easily seen that

$$\left.\begin{aligned} \left(x_2 \frac{\partial}{\partial x_3} - x_3 \frac{\partial}{\partial x_2}\right)(x_2^2 + x_3^2) &= 0 , \\ (\quad ,, \quad)\ (x_2 + i x_3) &= i (x_2 + i x_3) , \\ (\quad ,, \quad)\ (x_2 + i x_3)^n &= n i (x_2 + i x_3)^n . \end{aligned}\right\} \tag{21}$$

Thus $\left(x_2 \frac{\partial}{\partial x_3} - x_3 \frac{\partial}{\partial x_2}\right)$ has the eigenvalues $n i$ and the eigenfunctions $(x_2 + i x_3)^n$ belonging to these eigenvalues. But the eigenfunction $(x_2 + i x_3)^n$ does not satisfy the boundary conditions for wave functions. We therefore consider

$$\psi = (x_2 + i x_3)^n f(x_1, x_2^2 + x_3^2) . \tag{22}$$

This ψ is also an eigenfunction of $\left(x_2 \frac{\partial}{\partial x_3} - x_3 \frac{\partial}{\partial x_2} \right)$ belonging to the eigenvalue ni. One can choose f in such a manner that ψ does satisfy the boundary conditions. ψ is now a valid eigenfunction of m_1, belonging to the eigenvalue

$$m_1' = n\hbar ; \tag{23}$$

n can take all integral values.

We now consider the eigenvalue of $(m_1^2 + m_2^2 + m_3^2)$:

$$(m_1^2 + m_2^2 + m_3^2) = -\hbar^2 \Big\{ \left(x_2 \frac{\partial}{\partial x_3} - x_3 \frac{\partial}{\partial x_2} \right)^2 + \left(x_3 \frac{\partial}{\partial x_1} - x_1 \frac{\partial}{\partial x_3} \right)^2 + \left(x_1 \frac{\partial}{\partial x_2} - x_2 \frac{\partial}{\partial x_1} \right)^2 \Big\} . \tag{24}$$

It can be shown that this commutes with m_1, m_2, m_3. Also from potential theory one knows that the operator $\{ \cdot \cdot \cdot \}$ has the eigenfunctions $\psi = S_n(\theta, \varphi) f(r)$ belonging to the eigenvalues $\lambda = -n(n+1)$. Here r, θ, φ are polar co-ordinates and S_n is the spherical harmonic of order n. The eigenvalues of $m_1^2 + m_2^2 + m_3^2$ are thus $\hbar^2 n(n+1)$.
The magnitude of $\vec{m}$ can now be defined conveniently by

$$k = \sqrt{m_1^2 + m_2^2 + m_3^2 + \frac{1}{4}\hbar^2} - \frac{1}{2}\hbar \tag{25}$$

which has the eigenvalues $0, \hbar, 2\hbar, \ldots\ldots$

We can have a wave function that is a simultaneous eigen-function of k and a component of $\vec{m}$, m_1 say. The eigenvalues of m_1 for a given eigenvalue k' of k range from $-k'$ to $+k'$. This result also comes from the theory of spherical harmonics.

The angular momentum operators m_i possess the following remarkable properties

$$S m_i = m_i S \; , \quad i = 1, 2, 3, \tag{26}$$

where S is a scalar.

$$\left.\begin{aligned} m_1 A_1 - A_1 m_1 &= 0, \\ m_1 A_2 - A_2 m_1 &= i\hbar A_3, \\ m_1 A_3 - A_3 m_1 &= -i\hbar A_2, \end{aligned}\right\} \tag{27}$$

and six more relations obtained by cyclic permutations of indices 1,2,3, $\vec{A}$ being a vector. For the vectors $\vec{x}$, $\vec{p}$ one can directly verify (27). It is easily shown that if two vectors satisfy (27) then their vector product also satisfies them, and their scalar product satisfies (26). Thus by induction one can prove (26) and (27) for all scalars and vectors that can be constructed out of $\vec{x}$ and $\vec{p}$.

Lecture 8

The concept of angular momentum, which was introduced in the last lecture for the case of a single particle, can be directly extended for the case of a system of particles. If x_{1r}, x_{2r}, x_{3r} denote the coordinates and p_{1r}, p_{2r}, p_{3r} the components of momentum of the rth of a system of particles ($r = 1,2,3 \ldots\ldots n$), then the quantities

$$M_1 = \sum_r (x_{2r} p_{3r} - x_{3r} p_{2r})$$

with similar expressions for M_2, M_3 are the components of the total angular momentum.

The commutation relations for the angular momentum, given earlier for the case of a single particle, hold also for the case of the total angular momentum of a system of particles. These relations are :

(i) If S is any scalar, it commutes with the angular momentum components:

$$M_r S = S M_r \qquad = 1,2,3$$

(ii) If A_1, A_2, A_3 is any vector, then

$$M_1 A_1 - A_1 M_1 = 0,$$
$$M_1 A_2 - A_2 M_1 = i\hbar A_3,$$
$$M_1 A_3 - A_3 M_1 = -i\hbar A_2 .$$

To prove these relations one can firstly show that if we take the elementary vectors $\vec{x}$ or $\vec{p}$ for $\vec{A}$, the relations do hold. Next it can be shown that if two vectors satisfy the above commutation rules, their vector product also satisfies them;

and their scalar product satisfies the scalar relation given in (i). Similarly, if one has a vector satisfying relation (ii), its product with a scalar also satisfies the same. Thus by a process of induction, the commutation relations can be established for any quantity which can be built up of vectors and scalars. Similar commutation relations can be established for a tensor of any rank.

We observe that the commutation relations of the angular momentum components with any quantity depend only on the vector nature of the quantity. To trace the origin of this remarkable relation, we introduce the concept of rotation operators.

A rotation operator has the physical significance of changing a system in a well defined state into another state which is determined by the original state and the amount of the rotation together with the axis of rotation. It is a linear operator, and can easily be expressed if we work in the Schrödinger representation. For convenience in writing we consider the case of a single particle. The state of the particle will be represented by a Schrödinger wavefunction $\psi(x_1, x_2, x_3)$. On rotation through an angle θ around the x_3-axis, the state will be represented by a new wavefunction ψ^* which is given by

$$\psi^* = \psi(x_1 \cos\theta + x_2 \sin\theta, -x_1 \sin\theta + x_2 \cos\theta, x_3).$$

If we take θ to be small enough for squares and higher powers to be negligible,

$$\psi^* = \psi(x_1 + x_2\theta, -x_1\theta + x_2, x_3)$$

which on development by a Taylor's theorem at the point x_1, x_2, x_3 gives

$$\psi^* - \psi = \theta \left(x_2 \frac{\partial}{\partial x_1} - x_1 \frac{\partial}{\partial x_2} \right) \psi$$

The operator

$$R_3 = x_2 \frac{\partial}{\partial x_1} - x_1 \frac{\partial}{\partial x_2}$$

may be called the rotation operator about the x_3 -axis. For a system of particles the sum of operators of the type takes the place of the present expression.

In terms of R_3 , the angular momentum component M_3 may be written as

$$M_3 = \sum_r \left\{ x_{1r} \left(-i\hbar \frac{\partial}{\partial x_{2r}}\right) - x_{2r} \left(-i\hbar \frac{\partial}{\partial x_{1r}}\right) \right\}$$

$$= i\hbar R_3$$

and more generally

$$\vec{M} = i\hbar \vec{R}$$

where $\vec{M}$ denotes the vector M_1, M_2, M_3 and $\vec{R}$ the vector R_1, R_2, R_3

For an infinitesimal rotation θ , the operator giving the new state in terms of the old state is $1 + \theta R_3$; more generally for any rotation (finite or infinitesimal), the operator giving the new state is $e^{\theta R_3}$.

It is this relation existing between the angular momentum operator and the rotation operator that accounts for the remarkable commutation relations already encountered. They correspond to the law of transformation of a vector quality under a rotation.

The relation between angular momentum and rotation operators has been demonstrated above only for the case of a system of particles. However, angular momentum seems to be always connected with rotation in nature. We make the general assumption that the relation $\vec{M} = i\hbar \vec{R}$ always holds. In this way, we may introduce the angular momentum operator for the spin of a particle, even though no Schrödinger representation is possible. (See the end of the present lecture below).

Assuming this general relation to hold for the spin of a particle, and letting m_1, m_2, m_3 denote the components of the spin, we can infer the commutation relations

$$m_1 m_2 - m_2 m_1 = i\hbar\, m_3, \qquad \text{etc.}$$

We can use these relations to study the eigenvalues of m_1, m_2, m_3. The quantity $\beta = m_1^2 + m_2^2 + m_3^2$ is a scalar and hence commutes with m_1, m_2, m_3. We may take β to be a number without any inconsistency arising therefrom.

Let m_3' be an eigenvalue of m_3 and $|m_3'\rangle$ be the corresponding eigenket:

$$m_3 |m_3'\rangle = m_3' |m_3'\rangle .$$

Introduce a complex dynamical variable η defined by

$$\eta = m_1 - i m_2$$

so that

$$\bar{\eta} = m_1 + i m_2 .$$

Then

$$\begin{aligned} \bar{\eta}\eta &= (m_1 + i m_2)(m_1 - i m_2) = m_1^2 + m_2^2 - i(m_1 m_2 - m_2 m_1) \\ &= m_1^2 + m_2^2 + \hbar m_3 = \beta - (m_3 + \tfrac{1}{2}\hbar)^2 + \tfrac{1}{4}\hbar^2 \end{aligned}$$

and

$$\eta\bar{\eta} = \beta - (m_3 - \tfrac{1}{2}\hbar)^2 + \tfrac{1}{4}\hbar^2 .$$

We will use these results presently. In addition, we have the commutation relations

$$m_3\eta - \eta m_3 = m_3(m_1 - i m_2) - (m_1 - i m_2)\, m_3$$
$$= i\hbar m_2 - \hbar m_1 = -\hbar\eta .$$

Hence

$$m_3\, \eta\, |m_3'\rangle = (\eta m_3 - \hbar\eta)\; |m_3'\rangle$$
$$= (m_3' - \hbar)\; \eta\, |m_3'\rangle$$

Hence $\eta\, |m_3'\rangle$ is an eigenket corresponding to the eigenvalue $m_3' - \hbar$ provided $\eta\, |m_3'\rangle \neq 0$. For a minimum eigenvalue $m_3^{\min}$ to exist, we should have

$$\eta\, |m_3^{min}\rangle = 0$$

Taking the scalar product with the conjugate bra,

$$\langle m_3^{min}\, |\bar{\eta}\,\eta\, |\, m_3^{min}\rangle = 0$$

Using the known expression for $\bar{\eta}\,\eta$ (see above), this becomes

$$\langle m_3^{min}\, |\, \beta - (m_3^{min} - \tfrac{1}{2}\hbar)^2 + \tfrac{1}{4}\hbar^2\, |m_3^{min}\rangle = 0 .$$

From this it follows that

$$\beta - (m_3^{min} - \tfrac{1}{2}\hbar)^2 + \tfrac{1}{4}\hbar^2 = 0$$

so that

$$m_3^{min} = \tfrac{1}{2}\hbar \pm \sqrt{(\beta + \tfrac{1}{4}\hbar^2)} .$$

Similarly, it can be shown that $\bar{\eta}\, |m_3'\rangle$ is an eigenket belonging to the eigenvalue $m_3' + \hbar$ if $\bar{\eta}\, |m_3'\rangle \neq 0$. For a maximum eigenvalue $m_3^{\max}$ to exist, we should have

$$\bar{\eta}\,|m_3^{max}\rangle = 0$$

and we can show that

$$m_3^{max} = -\tfrac{1}{2}\hbar \pm \sqrt{(\beta + \tfrac{1}{4}\hbar^2)}\,.$$

Since $m_3^{max} \geq m_3^{min}$ we take the square roots in the expression for m_3^{min}, m_3^{max} in only one way and the ambiguity of the sign of the square root is removed:

$$m_3^{min} = \tfrac{1}{2}\hbar - \sqrt{(\beta + \tfrac{1}{4}\hbar^2)}\,,$$

$$m_3^{max} = -\tfrac{1}{2}\hbar + \sqrt{(\beta + \tfrac{1}{4}\hbar^2)}\,.$$

The difference between m_3^{max} and m_3^{min} should be an integral multiple of $\hbar$, so

$$2\sqrt{(\beta + \tfrac{1}{4}\hbar^2)} = (2k+1)\hbar$$

where k is an integer or half integer. The minimum possible value of k is zero. We may rewrite the relation in the form

$$\sqrt{(\beta + \tfrac{1}{4}\hbar^2)} - \tfrac{1}{2}\hbar = k\hbar \quad . \quad k \text{ integer or half integer.}$$

k so introduced can be conveniently defined to be the magnitude of the angular momentum; (it is in agreement with our previous definition) and we have

$$m_3' = -k\hbar\,,\ -k\hbar + \hbar\,,\ \ldots\,,\ +k\hbar\,.$$

Hence the eigenvalues of the spin angular momentum are all either integral or half integral. This is to be compared with the case of the orbital angular momentum (which could be represented in the Schrödinger formulation) where all the eigenvalues are integral. For the case of spin, the possibilities are more general than for orbital angular momentum.

The case of zero spin, i.e. $k = 0$ is a unique case, since all the components m_1, m_2, m_3 will have then zero eigenvalues. This possibility of being able to ascribe precise numerical magnitudes to noncommuting quantities simultaneously is unusual. It arises because of the peculiar commutation relation obeyed by the angular momentum components: the commutator of any two of the three quantities m_1, m_2, m_3 vanishes for the case of zero spin.

According to the results obtained above connecting angular momentum with rotation operator, the state of zero spin must be spherically symmetric, because it is unchanged by the application of a rotation. To illustrate this take the single case of a ruler placed on the table in a particular direction; here we can observe a definite orientation of the ruler in space and hence the 'state' of the ruler is not spherically symmetric. That a body can be in a non-symmetric state when it has no (macroscopic) angular momentum may be taken, at first sight, to be a contradiction to the general result enunciated above. However, at any finite temperature, the body is subject to thermal oscillations and the number of elementary quanta of angular momentum associated with this is very large. If one could cool the ruler to such a temperature that no quanta are associated with it, the state will have to become spherically symmetric. That is to say, any experiment designed to give the direction of the length of the ruler would give all direction with equal probability. Another example is the electron wavefunction in the ground state of the hydrogen atom neglecting spins; the angular momentum is zero and the state is spherically symmetric.

We will now consider the case of spin $\frac{1}{2}$. Here $k = \frac{1}{2}\hbar$, $m_1' = \pm\frac{1}{2}\hbar$ and similarly for m_2' and m_3'. $m_1^2 = \frac{1}{4}\hbar^2$ always. Since it has only one eigenvalue we may take it to be a number. One can now introduce the σ variables with advantage by putting

$$m_r = \frac{1}{2}\hbar\ \sigma_r\ , \qquad 1,2,3.$$

Then

$$\sigma_r^2 = 1$$

and the commutation relation

$$\sigma_1\sigma_2 - \sigma_2\sigma_1 = 2i\sigma_3 \tag{1}$$

and other relations obtained by the cyclic permutation of the indices 1,2,3 will be valid. These relations can be simplified by multiplying. We multiply (1) by σ_1 first on the left, then on the right and add

$$\sigma_1(\sigma_1\sigma_2 - \sigma_2\sigma_1) + (\sigma_1\sigma_2 - \sigma_2\sigma_1)\sigma_1 = 2i(\sigma_1\sigma_3 + \sigma_3\sigma_1).$$

Remembering that $\sigma_r^2 = 1$ it follows that

$$\sigma_2 - \sigma_1\sigma_2\sigma_1 + \sigma_1\sigma_2\sigma_1 - \sigma_2 = 2i(\sigma_1\sigma_3 + \sigma_3\sigma_1),$$

giving

$$\sigma_1\sigma_3 + \sigma_3\sigma_1 = 0.$$

A set of three 2 x 2 matrices satisfying the relations $\sigma_r^2 = 1$, $\sigma_1\sigma_2 = -\sigma_2\sigma_1 = i\sigma_3$ were first given by Pauli; and are known as Pauli matrices.

These relations satisfied by the σ variables are identical with those satisfied by the nonscalar unit elements of quaternions*, apart from a factor i. In a previous lecture we studied quantities satisfying these relations and obtained matrices to represent them (Pauli's matrices).

A general remark may be made about the spin eigenvalues and the possibility of a Schrödinger representation. From the relation $m_3 = i\hbar R_3$ for an eigenstate of m_3 belonging to the eigenvalue m_3' we have

$$R = \frac{m_3'}{i\hbar}$$

Hence the operator for a finite rotation is

$$e^{\theta R_3} = e^{-i\theta m_3'/\hbar}$$

For a complete rotation $\theta = 2\pi$ and we have

$$e^{\theta R_3} \longrightarrow e^{-2\pi i m_3'/\hbar}$$

---

* The algebra of quaternions (invented by W.R. Hamilton) is generated by the four elements e_0, e_1, e_2, e_3 which satisfy the relations,

$$e_0^2 = -e_1^2 = -e_2^2 = -e_3^2 = e_0,$$

$$e_0 e_1 = e_1 e_0 = e_1,$$

$$e_1 e_2 = -e_2 e_1 = e_3,$$

and similar relations obtained by cyclic permutation of the indices 1,2,3.

If $m_3' = \frac{1}{2}\hbar$ this becomes $e^{-\pi i} = -1$. Hence a ket corresponding to a spin ½ state would change sign on rotating through 2π radians. The same is true of any half integral spin state. For any Schrödinger function, a rotation through 2π must leave the function unchanged, we see that the Schrödinger representation is possible only for the case of integral spins (in which case the finite rotation operator $e^{\theta R_3}$ becomes + 1 for $\theta = 2\pi$).

Lecture 9

It was shown in the last lecture that a component of the spin angular momentum of a particle can have only integral or half-odd-integral eigenvalues (in units of $\hbar$). The inadmissibility of arbitrary fractions of for the values of the spin is a remarkable fact and can also be understood from arguments of a general character.

Consider a rigid body capable of being rotated about a fixed axis. The orientation of the body can be determined by three variables (the Euler angles, say). We represent a given orientation of the body by a point in a certain three dimensional space called the orientation space. We wish to consider the topological properties of this space.

If the rigid body be swung to some other orientation, the point representing its orientation will move along some path in the orientation space; if the body be brought back to its original orientation, this path will become a closed loop. If the body were swung in a slightly different manner to get back to its original orientation, we will obtain a slightly different closed loop. The question we wish to consider is whether the manner of swinging can be changed in a continuous way so as to coalesce the loop to a point.

A single revolution of the body about a fixed axis is represented by a loop which cannot be continuously shrunk to a point. Two revolutions about the same axis will be represented by the same loop traversed twice. A simple proof has been found by Tania Ehrenfest to show that this double loop corresponding to two revolutions can be shrunk to a point in a continuous manner.

Consider two rigid cones 1 and 2 of the same angle α and with

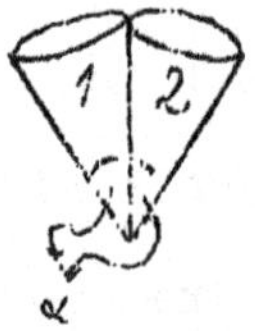

their vertices coinciding and suppose cone 2 to roll round cone 1 without slipping as represented in the figure. Cone number 2 is our rotating rigid body, and its going around the other completely once will be represented by some closed loop in the orientation space, since the final orientation of the cone 2 is evidently the same as the original orientation. Now let α vary from 0 to π. For $\alpha = 0$ the cones become two cylinders, and the motion of the moving cone becomes two revolutions about the axis of the fixed cone. For α very near to π one sees that the motion becomes just a slight wobble, and in the limit $\alpha \to \pi$ this too will disappear, and we have no motion at all. In this way the loop corresponding to two revolutions can be continuously shrunk to zero.

Now suppose that a body could have spin $\frac{1}{3}\hbar$. The ket vector representing a given state of it will get multiplied by a complex cube root of unity, ω say, on making a rotation about an axis. Two rotations about the same axis will cause the ket to be multiplied by ω^2, which is also a cube root of unity. Now the cube root of unity cannot be reduced to unity continuously, whereas a double rotation can be reduced continuously to no motion at all, so that the ket vector obtained on making two revolutions must be the original ket and the spin of $\frac{1}{3}\hbar$ cannot occur. It is this kind of contradiction which rules out all but the integral and half-odd-integral values for the spin of a particle.

To describe half-odd-integral spins one uses certain quantities called spinors which have the property of changing sign on making a revolution about an axis. A mathematical theory has been built up about them, called spinor analysis. The elementary particles met with so far in nature have spin values 0, ½, and 1. For particles of spin ½ the theory is sufficiently simple for one to do without spinor analysis, and it is not necessary to deal with it here.

We shall now consider the spin of the photon. We shall first find how to define the spin of a particle in general. The total angular momentum is a well defined quantity, because of its connection with the rotation operator. We put

$$\vec{M} = \vec{m}_{orb} + \vec{m}_{spin}$$

The question is how to separate $\vec{m}_{spin}$ and $\vec{m}_{orb}$. Now the translation operators are well defined and connected with the total linear momentum in a similar way to that in which the rotation operators are connected with the total angular momentum. Thus the total linear momentum p_r $(r = 1, 2, 3)$ is well-defined. The variables x_r can now be defined as the conjugates of the p_r, but this does not determine them completely. One could, without violating the quantum conditions $x_r p_s - p_s x_r = i\hbar\delta_{rs}$, replace x_r by

$$x_r^* = x_r + \xi_r,$$

ξ_r being a function of the p's and the internal variables, so that ξ_r commutes with the p's; x_r^*, then also satisfies

$$x_r^* p_s - p_s x_r^* = i\hbar\delta_{rs}$$

and provided the ξ 's satisfy suitable commutation relation to make the x^* 's commute, the x^* 's are just as good conjugates to the p 's as the x 's are.

We thus obtain the following two expressions for the spin angular momentum, which are equally good.

$$m_{spin} = M - x \times p \, ,$$

$$m^*_{spin} = M - x^* \times p \, .$$

Their difference

$$m_{spin} - m^*_{spin} = \xi \times p$$

gives us the ambiguity in the definition of spin angular momentum.

For a state of the system in which the momentum p is zero, $\xi \times p$ will be zero, since ξ and p commute, and the ambiguity in the definition of m_{spin} disappears. The spin angular momentum is thus uniquely defined for a system with linear momentum equal to zero.

For a photon this method is not applicable since a photon moves with the speed c always and its momentum cannot be zero. Consider a photon moving in the direction of the x_3 axis. Then

$$p_1 = p_2 = 0$$

and we can define the component $m_{\text{spin}\,3}$ in an unambiguous manner following the procedure of the preceding paragraph. Thus one component of the spin of a photon is well defined, namely the component about an axis in the direction of motion; the other two components cannot be defined. If a state of a photon moving in the direction x_3 is an eigen ket of m_3 belonging to the eigenvalue m_3' then on making a rotation through an angle θ about the x_3 axis this ket will get multiplied by the number $e^{i m_3' \theta / \hbar}$. The state is thus not changed. Hence eigenkets of m_3 will be states of circular polarization. States of linear polarization are obtained by taking linear combinations of the two states of circular polarization in opposite senses, with an equal weight for each, and a phase relationship depending on the direction of the linear polarization. It can be shown that the spin of the photon about its direction of motion is plus or minus one quantum. Let the kets $|1\rangle$ and $|2\rangle$ represent the states of linear polarization in two perpendicular directions. Consider a rotation through an angle θ. This is produced by the operator $e^{R\theta}$, R being the infinitesimal rotation operator. We have

$$e^{R\theta}|1\rangle = |1\rangle\cos\theta + |2\rangle\sin\theta$$

For small θ

$$R|1\rangle = |2\rangle,$$

and similarly

$$R|2\rangle = -|1\rangle.$$

The angular momentum operator is $m = i\hbar R$. We have

$$m\{|1\rangle \pm i|2\rangle\} = i\hbar R\{|1\rangle \pm i|2\rangle\}$$

$$= \pm\hbar\{|1\rangle \pm i|2\rangle\}$$

Thus m has eigenvalues $\hbar$, $-\hbar$ and the photon has a spin $\pm\hbar$ about its direction of motion.

One cannot give a meaning to its spin in other direction, so one cannot give a meaning to k, the magnitude of the spin vector.

In classical mechanics a state of a system of particles can be represented by a point in the phase space. When the state of the system is not completely known, then the state is represented, according to the method of Gibbs, by a cloud of points with density $\rho(p, q)$; $\rho(p,q)\,dp\,dq$ is the probability for the system to be in the state defined by the momenta and coordinates being between p , $p+dp$ and q , $q+dq$ respectively. The average value of a dynamical variable is then given by

$$\xi_{av} = \int \rho \xi \, dp \, dq. \tag{1}$$

Although the concept of phase space has no meaning in quantum mechanics, we can define a linear operator ρ corresponding to the Gibbs density ρ . Suppose that a system is distributed over various states, represented by the normalized kets

$$|x_1\rangle, |x_2\rangle, \ldots \tag{2}$$

with probabilities

$$P_1, P_2, \ldots \tag{3}$$

respectively. We define the linear operator

$$\rho = \sum_n |x_n\rangle P_n \langle x_n| \,. \tag{4}$$

ρ defined in this way has properties analogous to the Gibbs density ρ .

The average value of an observable ξ for the ensemble defined by (2) and (3) is

$$\xi_{av} = \sum_n P_n \langle x_n | \xi | x_n \rangle \,.$$

If $|r\rangle$ is a base of coordinates so that

$$\sum_r |r\rangle\langle r| = 1,$$

then

$$\xi_{av} = \sum_r \sum_n P_n \langle x_n | \xi | r \rangle \langle r | x_n \rangle$$

$$= \sum_r \sum_n \langle r | x_n \rangle P_n \langle x_n | \xi | r \rangle$$

$$= \sum_r \langle r | \rho \xi | r \rangle$$

$$= \sum_r \langle r | \xi \rho | r \rangle . \qquad (5)$$

Thus ρ allows one to calculate the avërage value of observables. The summing of the diagonal elements of $\rho \xi$ corresponds to the integration over phase space in (1) of classical mechanics.

When nothing at all is known about a system, it could with equal likelihood be in any state, so we take a set of orthogonal kets $|x_n\rangle$ for (2) and we take all the P_n equal to $\frac{1}{N}$ in (3), N being the number of independent states. Then ρ defined by (4) is just the number $\frac{1}{N}$.

Thus for completely unpolarized light we have

$$\rho = \tfrac{1}{2} |1\rangle\langle 1| + \tfrac{1}{2} |2\rangle\langle 2| = \tfrac{1}{2}$$

where $|1\rangle$ and $|2\rangle$ are the normalized kets representing two states of linear polarization in two perpendicular directions.

Lecture 10

Consider a set of quantities (linear operators) $\alpha, \beta, \gamma, \ldots$ and the transformation

$$\alpha \longrightarrow \alpha^* = S\alpha S^{-1}$$

acting on them, where S is any operator for which a reciprocal exists. It is easily seen that the transformed quantities $\alpha^*, \beta^*, \ldots$ satisfy the same algebraic relations as the original quantities $\alpha, \beta, \ldots$ since

$$\alpha^* + \beta^* = S\alpha S^{-1} + S\beta S^{-1} = S(\alpha+\beta)S^{-1}$$

so that if $\alpha+\beta=\gamma$, then $\alpha^*+\beta^*=\gamma^*$ and similarly

$$\alpha^*\beta^* = S\alpha S^{-1}S\beta S^{-1} = S\alpha\beta S^{-1}$$

so that if $\alpha\beta=\gamma$ then $\alpha^*\beta^*=\gamma^*$. If in addition we define the corresponding transformations of bra and ket vectors by the relations

$$|P^*\rangle = S|P\rangle$$
$$\langle P^*| = \langle P|S^{-1}$$

it is seen that all relations involving operators, bras and kets are formally preserved by the transformation.

We now impose two conditions on the transformations (1) α^* should be real if α is real; (2) $\langle P^*|$ should be the conjugate of $|P^*\rangle$. The second condition means that

$$\langle P|S^{-1} = \text{Conjugate of } S|P\rangle = \langle P|\bar{S},$$

and since this is to be true for all $\langle P|$, it follows that

$$S^{-1} = \bar{S} \quad \text{or} \quad S\bar{S} = 1.$$

If S satisfies this relation the first condition is also satisfied, since

$$\text{conjugate of } \alpha^* = \overline{S\alpha\bar{S}} = \bar{\bar{S}}\ \bar{\alpha}\ \bar{S} = S\alpha\ \bar{S}$$
$$= \alpha^*,$$

(since $\bar{\bar{S}} = S$ and $\bar{\alpha} = \alpha$ since α is real) showing that α^* is real. A linear operator S which satisfies the relation

$$\bar{S}S = 1$$

is called unitary and the transformation induced by S a unitary transformation. We see that the new quantities (linear operators, bras and kets) got by a unitary transformation are just as good as the original quantities for the description of the system, since they satisfy the same algebraic relations.

So far our discussions have been restricted to describing the condition of a system at one instant of time. Such a discussion, while trivial in classical mechanics, was undertaken to aid us in the understanding of the basic principles of quantum theory. We must now deal with the change in quantum mechanical systems with time and set up equations of motion. For this purpose some new physical assumptions must be made.

Denote a state of the system at time t_0 by $|A t_0\rangle$ at a later time t it will go over into a different state $|At\rangle$. We now make the fundamental assumption that $|At\rangle$ is determined by $|At_0\rangle$, so that causality holds, provided no disturbance of the system, of the nature of an observation, is made during the time interval t_0 to t. We make the further fundamental assumption that the superposition principle holds for the states of the system at <u>any time</u> This means that if there is a superposition relation between certain states initially, this superposition relation will be preserved, so long as the system is undisturbed by observations.

This assumption requires that the ket $|At\rangle$ corresponding to time t should depend linearly on the ket $|At_0\rangle$ at time t_0

$$|At\rangle = T\,|At_0\rangle$$

where T is a linear operator, which is the same for all the states

In addition, we assume that T preserves the length of the kets:

$$\langle At|At\rangle = \langle At_0|At_0\rangle .$$

This assumption is sufficient to ensure that scalar products are preserved; to see this, apply the previous equation to the kets $|A\rangle + |B\rangle$ and $|A\rangle + i\,|B\rangle$ where $|A\rangle$ and $|B\rangle$ are any two kets. This gives

$$\left(\langle At| + \langle Bt|\right)\left(|At\rangle + |Bt\rangle\right)$$
$$= \langle At|At\rangle + \langle At|Bt\rangle + \langle Bt|At\rangle + \langle Bt|Bt\rangle$$

and $(\langle At| - i\langle Bt|)(|At\rangle + i|Bt\rangle)$

$$= \langle At|At\rangle + i\langle At|Bt\rangle - i\langle Bt|At\rangle + \langle Bt|Bt\rangle$$

are independent of t. But $\langle At|At\rangle$ and $\langle Bt|Bt\rangle$ are independent of time. Hence it follows that $\langle At|Bt\rangle$ and $\langle Bt|At\rangle$ are also independent of time. Hence

$$\langle At_0|\bar{T}T|Bt_0\rangle$$

is independent of time for all $|At_0\rangle$ and $|Bt_0\rangle$

It follows that

$$\bar{T}T = 1$$

and T is a unitary operator.

For mathematical convenience we take the time interval $t - t_0$ small and see what the relation between $|At\rangle$ and $|At_0\rangle$ then becomes. We appeal to a principle of physical continuity, that when $t - t_0$ tends to zero, the change in the ket also tends to zero, so that the operator T must tend to unity:

$$t \to t_0 , \qquad T \to 1$$

Form the quantity

$$\frac{\partial |At\rangle}{\partial t} = \lim_{t \to t_0} \left\{ \frac{|At\rangle - |At_0\rangle}{(t - t_0)} \right\}$$

$$= \lim_{t \to t_0} \frac{T-1}{t-t_0} |At_0\rangle$$

We assume that the limit exists. This limit is also a linear operator. From the unitary condition it is pure imaginary as is shown below:

If S is a unitary operator and given by

$$1 + \epsilon X$$

where ε is an infinitesimal real number. Then $\bar{S} = 1 + \varepsilon \bar{X}$ and making use of the unitary property $\bar{S} S = 1$

$$\varepsilon (X + \bar{X}) = 0$$

so X is pure imaginary.

Our present T is of the form $S = 1 + \varepsilon X$ with $\varepsilon = t - t_0$ and hence

$$\lim_{t \to t_0} \frac{T - 1}{t - t_0}$$

is pure imaginary. The relation between $\frac{\partial |A\rangle}{\partial t}$ and $|A t_0\rangle$ is usually given in the form

$$i\hbar \frac{d |A\rangle}{dt} = H |A\rangle$$

where H is now a <u>real</u> operator (by virtue of the i on the left side of the equation).

The operator H determines the motion of the system completely, and is called the Hamiltonian operator. The above equation is Schrödinger's equation of motion.

Lecture 11

In the last lecture it was shown that the state $|Pt\rangle$ of a system left undisturbed satisfies the differential equation

$$i\hbar \frac{d}{dt} |Pt\rangle = H |Pt\rangle$$

and that the kets corresponding to the states of the system at different times are connected by a unitary transformation:

$$|Pt\rangle = T |Pt_0\rangle$$

The dynamical variables in this picture are constant linear operators.

We can get another picture of the dynamical system by performing a (time-dependent) unitary transformation on all the quantities (both kets and operators), such that the representative ket is brought to rest, i.e., is constant in time. If α is transformed into α^* such that

$$\alpha^* = T^{-1} \alpha T$$

the corresponding transforms of ket $|x\rangle$ and bra $\langle x|$ are

$$|x\rangle^* = T^{-1} |x\rangle, \quad \langle x|^* = \langle x| T.$$

This makes $|Pt\rangle^*$ constant, i.e. $|Pt\rangle^* = |Pt_0\rangle^*$

The dynamical variables α^* are then moving. One has

$$\frac{d\alpha^*}{dt} = \frac{dT^{-1}}{dt} \alpha T + T^{-1} \alpha \frac{dT}{dt}.$$

Since $|Pt\rangle = T\,|Pt_0\rangle$ we get on differentiation

$$H\,|Pt\rangle = i\hbar \frac{d}{dt}|Pt\rangle = i\hbar \frac{dT}{dt}|Pt_0\rangle$$

Observing that the first member of the equation can be written in the form

$$H\,T\,|Pt_0\rangle$$

and since the relation holds for all $|Pt_0\rangle$ one obtains

$$i\hbar \frac{dT}{dt} = H\,T.$$

Also $\dfrac{dT^{-1}}{dt} = -\dfrac{1}{T}\dfrac{dT}{dt}\dfrac{1}{T}$ (since this make $\dfrac{d}{dt}(T\,T^{-1}) = 0$)

so that

$$i\hbar \frac{dT^{-1}}{dt} = -\frac{1}{T}\,H.$$

Making use of these equations the equation for $\dfrac{d\alpha^*}{dt}$ can be written as

$$i\hbar \frac{d\alpha^*}{dt} = -\frac{1}{T}\,H\,\alpha\,T + \frac{1}{T}\,\alpha\,H\,T.$$

Putting $H^* = T^{-1}\,H\,T$

we have hence the equation of motion for

$$i\hbar \frac{d\alpha^*}{dt} = \alpha^* H^* - H^* \alpha^*.$$

This is called Heisenberg's equation of motion.

We have thus two schemes of quantum mechanics which are connected by a (time dependent) unitary transformation of the various quantities. The relationship between the two schemes may best be displayed in the form of a table.

Scheme	States	Dyanamical Variables
Schrödinger representation	moving	fixed
Heisenberg representation	fixed	moving

Both the schemes were discovered independently, Heisenberg's discovery preceding Schrodinger's by a few months. They are mathematically equivalent, since they are connected by a unitary transformation.

Assessing relative merits, the Schrodinger scheme is better suited to actual calculation since the (unknown) variable is a ket in this scheme; while in the other scheme, the (unknown) variable is a linear operator. For purposes of calculation a ket is far simpler to deal with than a linear operator.

The Heisenberg scheme, on the other hand, is in direct correspondence with classical mechanics where the dynamical variables change with time. For the comparison between the Heisenberg scheme and classical mechanics, the best method is to put classical mechanics into the Hamiltonian form. When the Lagrange function of a system is given, one can, by standard methods, pass on to the Hamiltonian equations of motion of classical mechanics:

$$\frac{dq_r}{dt} = \frac{\partial H}{\partial p_r}, \qquad \frac{dp_r}{dt} = -\frac{\partial H}{\partial q_r},$$

where $H = H(p, q)$ is the Hamiltonian function, which is the total energy, expressed in terms of the canonical variables p, q. If v is any dynamical quantity which is expressible as a function of q, p one has

$$\frac{dv}{dt} = \sum_r \left(\frac{\partial v}{\partial q_r} \frac{dq_r}{dt} + \frac{\partial v}{\partial p_r} \frac{dp_r}{dt} \right)$$

$$= \sum_r \left(\frac{\partial v}{\partial q_r} \frac{\partial H}{\partial p_r} - \frac{\partial v}{\partial p_r} \frac{\partial H}{\partial q_r} \right)$$

which may be written in the form

$$\frac{dv}{dt} = [v, H]$$

in Poisson bracket notation. A Poisson bracket is defined in classical mechanics by

$$[\alpha, \beta] = \sum_r \left(\frac{\partial \alpha}{\partial q_r} \frac{\partial \beta}{\partial p_r} - \frac{\partial \alpha}{\partial p_r} \frac{\partial \beta}{\partial q_r} \right).$$

There is a corresponding Poisson bracket in quantum theory, defined by

$$[\alpha, \beta] = \frac{\alpha\beta - \beta\alpha}{i\hbar}$$

In either case we have the following relations holding:

$$[\alpha, \beta] = -[\beta, \alpha] ,$$

$$[\alpha_1 + \alpha_2, \beta] = [\alpha_1, \beta] + [\alpha_2, \beta] ,$$

$$[\alpha_1 \alpha_2, \beta] = \alpha_1 [\alpha_2, \beta] + [\alpha_1, \beta] \alpha_2 ,$$

$$[[\alpha, \beta], \gamma] + [[\beta, \gamma], \alpha] + [[\gamma, \alpha], \beta] = 0 .$$

Further, choosing α, β to be canonical coordinates and momenta, we obtain for the Poisson brackets both in classical theory as well as in quantum theory.

$$[q_r, q_s] = 0,$$
$$[p_r, p_s] = 0,$$
$$[q_r, p_s] = \delta_{rs}.$$

In view of this formal similarity, one can consider the quantum Poisson brackets to replace the classical Poisson brackets in the transition from classical mechanics to quantum theory.

Making use of the quantum Poisson brackets one can express Heisenberg's equation of motion in the form

$$\frac{d\alpha^*}{dt} = [\alpha^*, H^*]$$

which is completely identical formally with the classical relation

$$\frac{dv}{dt} = [v, H].$$

It is because of this analogy that H is called the Hamiltonian in quantum theory. Since H represents the total energy in the classical case, we assume that the same holds in quantum theory also.

It is seen that the formal structure of quantum theory is intimately connected with that of classical mechanics. This makes it seem impossible to change quantum mechanics in any way without spoiling the entire scheme. To make progress at the present day, instead of trying any such changes one should concentrate on obtaining better Hamiltonians. The difficulties encountered in present day theories of elementary particles may be entirely due to people working with the wrong Hamiltonians.

The Hamiltonian chosen in any classical case may be either a constant or dependent on time according to whether we are considering a "free" system or a system acted upon by external forces. The same choice may be made in quantum mechanics.

In talking about whether a Hamiltonian is constant or not, it is immaterial to specify whether we are dealing with the Schrödinger or the Heisenberg scheme, since if the Hamiltonian is constant in one scheme, it can be shown to be constant in the other also. For example, if H is constant (i.e., independent of t) in the Schrodinger picture, we can put $\alpha = H$ in the Heisenberg equation of motion obtaining

$$i\hbar \frac{dH^*}{dt} = H^*H^* - H^*H^* = 0 ,$$

so that H^* is a constant in the Heisenberg picture.

If we are dealing with a closed system, the total Hamiltonian is a constant. A time-dependent Hamiltonian presupposes the operation of external forces and the reaction on these agencies by the system considered is neglected. This procedure is an approximation which is useful in some cases, but cannot provide an exact theory.

For a stationary state in the Schrodinger picture, the direction of the ket vector is unaltered and hence

$$\frac{d}{dt} |Pt\rangle = \lambda |Pt\rangle$$

where λ is a number. Comparing with Schrodinger's equation of motion

$$i\hbar \frac{d}{dt} |Pt\rangle = H |Pt\rangle$$

one obtains

$$i\hbar\lambda\,|Pt\rangle = H\,|Pt\rangle .$$

Hence the stationary states are the eigenstates of the Hamiltonian:

$$i\hbar\lambda = H'$$

where H' is an eigenvalue of H . We have then

$$\frac{d}{dt}\,|Pt\rangle = \frac{H'}{i\hbar}\,|Pt\rangle$$

giving an integration

$$|Pt\rangle = e^{-i\frac{H't}{\hbar}}\,|Pt_0\rangle$$

Hence the ket varies periodically with a frequency depending upon H.
The energy is assumed to be an observable, so there exists a sufficient number of stationary states for any state to be expressed as a linear combination of them.

If we suppose that H is a constant, the Schrödinger equation of motion can be solved to obtain an integral

$$|Pt\rangle = e^{-iH(t-t_0)/\hbar}\,|Pt_0\rangle .$$

That this solution satisfies the equation of motion can be verified by direct substitution.* The corresponding solution of the Heisenberg

---

* In differentiating, ordinary algebra can be used here, since no noncommuting pair of variables appear in the process.

equation of motion is

$$\alpha_t^* = e^{iH(t-t_0)/\hbar} \alpha_{t_0}^* e^{-iH(t-t_0)/\hbar}$$

The solutionsso obtained are not usually of much utility in specific problems due to the difficulty of computing the exponential of a linear operator.

Transition Theory: Consider a complete set of commuting observables α and for convenience assume the eigenvalues of α to be discrete. Adopting the Schrödinger picture, we will have a time dependent wave function $\psi(\alpha, t)$ and the equations of motion would be given by,

$$i\hbar \frac{d}{dt} \psi(\alpha,t) \, |S\rangle = H \, \psi(\alpha,t) \, |S\rangle$$

so that

$$i\hbar \frac{d}{dt} \psi(\alpha,t) = H \, \psi(\alpha,t)$$

The probability of a value $\alpha = \alpha'$ at time t is $|\psi(\alpha',t)|^2$ provided $\psi(\alpha)$ is normalized. This result will be put in a different form, to bring out a certain symmetry.

Introduce simultaneous eigenkets of all the α 's and let these be denoted by $|\alpha'\rangle = \delta_{\alpha\alpha'} \, |S\rangle$ in standard ket notation. For any general ket

$$|P\rangle = \psi(\alpha) \, |S\rangle$$

we can form the product

$$\langle \alpha' | P \rangle = \langle S | \delta_{\alpha\alpha'} \psi(\alpha) | S \rangle$$
$$= \sum_{\alpha} \delta_{\alpha\alpha'} \psi(\alpha) = \psi(\alpha').$$

Hence the probability of a value $\alpha = \alpha'$ can be expressed in the form

$$|\langle \alpha' | P \rangle|^2$$

The probability of a value $\alpha = \alpha'$ at a time t is now

$$|\langle \alpha' | Pt \rangle|^2 = |\langle \alpha' | T | Pt_0 \rangle|^2$$

Suppose the initial state also is an eigenstate of α belonging to the eigenvalue α° :

$$|Pt_0\rangle = |\alpha^{\circ}\rangle$$

Then the probability of a transition from the value α° at t_0 to a value α' at time t can be expressed in the symmetrical form

$$|\langle \alpha' | T | \alpha^{\circ} \rangle|^2$$

The transition probability is thus expressed as the square of a matrix element. The (complex) matrix element $\langle \alpha' | T | \alpha^{\circ} \rangle$ may be called a probability amplitude. One must take the square of its modulus to get a probability.

A constant of the motion is a quantity which has zero time derivative in the Heisenberg picture: i.e. if α is a constant of motion,

$$\frac{d\alpha}{dt} = 0 \quad \text{or} \quad [\alpha, H] = 0 .$$

The above calculation of a transition probability is useful only if the α's, though not exact constants of motion, are very nearly so, so that the transition probability is small.

Lecture 12

It has been shown that if the kets are normalised, the probability of transition from the state α^0 at time t_0 to state α' at time t' is $|\langle\alpha'|T|\alpha^0\rangle|^2$ where $\langle\alpha^0|$ and $\langle\alpha'|$ are both discrete eigenstates. Similar results hold even if either or both states belong to a continuous set. If $\langle\alpha^0|$ is a discrete state and $\langle\alpha'|$ belongs to a continuous set $|\langle\alpha'|T|\alpha^0\rangle|^2$ is the probability of transition from α^0 to α' per unit range of α'. If α^0 is itself belonging to a continuous set the ket $|\alpha^0\rangle$ cannot be normalised. In this case one can only say that $|\langle\alpha'|T|\alpha^0\rangle|^2$ is proportional to the probability of the transition.

Consider a system whose Hamiltonian can be written down in the form

$$H = E + V$$

where E is a simple operator and the additional term V is small. As an example, E might be the proper energy of two things, i.e. the sum of their energies alone, and V the energy of interaction between them.

Choose the α's to be a complete set of commuting observables for a system whose Hamiltonian is E (and not $E+V$). For this system the eigenvalues of α will be constants of the motion, since these α commute with E. This system will be called the "unperturbed" system.

Adopt the Schrödinger picture where the ket $|P_t\rangle$ at any time t is given by

$$|P_t\rangle = T\,|P_{t_0}\rangle$$

and defines two new quantities T^*, $|P_t\rangle^*$ by the relations

$$T^* = e^{iE(t-t_0)/\hbar}\,T$$

$$|P_t\rangle^* = T^*\,|P_{t_0}\rangle = e^{iE(t-t_0)/\hbar}\,|Pt\rangle$$

Making use of Schrödinger's equation of motion

$$i\hbar \frac{d}{dt}|Pt\rangle = H\,|Pt\rangle$$

one obtains the equation of motion for $|Pt\rangle^*$;

$$i\hbar \frac{d}{dt}|Pt\rangle^* = e^{iE(t-t_0)/\hbar}\left\{-E\,|Pt\rangle + i\hbar\frac{d}{dt}|Pt\rangle\right\}$$

$$= e^{iE(t-t_0)/\hbar}\,V\,|Pt\rangle = V^*|Pt\rangle^*$$

where

$$V^* = e^{iE(t-t_0)/\hbar}\,V\,e^{-iE(t-t_0)/\hbar}$$

Thus the ket $|Pt\rangle^*$ satisfies an equation similar to the one satisfied by $|Pt\rangle$ but involving only V^*. We have a corresponding equation of motion for T^*;

$$i\hbar \frac{d}{dt}T^* = V^*T^*$$

The transition probability $|\langle\alpha'|T|\alpha^\circ\rangle|^2$ may be written in the form $|\langle\alpha'|e^{-iE(t-t_0)/\hbar}\,T^*|\alpha^\circ\rangle|^2$. Now $\langle\alpha'|$ is an eigenbra of the operator E so that

$$\langle\alpha'|E = E'\langle\alpha'| \,,\quad E' = E(\alpha')$$

and hence

$$\langle \alpha' | \, e^{-i E(t-t_0)/\hbar} = e^{-i E'(t-t_0)/\hbar} \langle \alpha' | ,$$

so $\langle \alpha' | T | \alpha^0 \rangle = e^{-i E'(t-t_0)/\hbar} \langle \alpha' | T^* | \alpha^0 \rangle$.

The coefficient of $\langle \alpha' | T^* | \alpha^0 \rangle$ on the right handside is a number of unit modulus, so when the modulus is taken, it can be suppressed. Hence the expression for the transition probability reduces to

$$| \langle \alpha' | T^* | \alpha^0 \rangle |^2$$

Thus in calculating the probability amplitudes, one might use T^* in place of T.

It is to be noted that the starred quantities T^*, $|Pt\rangle^*$ vary only slowly with time and this variation is solely due to the perturbation term V in the Hamiltonian. This picture is called the "Interaction Representation" and is employed in modern developments in Field Theory. On contrasting the slow variation of the kets in this representation with the rapid variation in the Schrödinger representation and the time-independent kets of the Heisenberg representation, one may say that the Interaction representation lies between the Heisenberg and Schrödinger representations.

All the developments so far have been exact and no approximations have been made. Now we proceed to consider a method of approximation generally used for calculating transition probabilities. T^* satisfies the differential equation

$$i\hbar \frac{d}{dt} T^* = V^* T^*$$

with the initial condition

$$T^* = 1 \quad \text{for} \quad t = t_0 .$$

Express T^* in the form

$$T^* = 1 + T_1^* + T_2^* + T_3^* + \cdots,$$

where the successive terms are of the order of ascending powers of V. Then we have the relations

$$i\hbar \frac{dT_1^*}{dt} = V^*,$$

$$i\hbar \frac{dT_2^*}{dt} = V^* T_1^*,$$

$$\cdots$$

$$i\hbar \frac{dT_n^*}{dt} = V^* T_{n-1}^*.$$

They have the solutions

$$T_1^*(t_1) = \frac{1}{i\hbar} \int_{t_0}^{t_1} V^*(t)\, dt,$$

$$T_2^*(t_2) = \frac{1}{i\hbar} \int_{t_0}^{t_2} V^*(t_1)\, dt_1 \int_{t_0}^{t_1} V^*(t)\, dt,$$

and similar iterated expressions for the higher order terms.

In a first approximation, the transition probability is given by

$$P(\alpha^0 \alpha') = |\langle\alpha'| T^* |\alpha^0\rangle|^2 = \frac{1}{\hbar^2} \left|\langle\alpha'| \int_{t_0}^{t_1} V^*(t)\, dt\, |\alpha^0\rangle\right|^2$$

$$\alpha' \neq \alpha^0$$

which can, by performing the integration after the matrix element has been taken, be expressed in the form

$$\frac{1}{\hbar^2} = \left| \int_{t_0}^{t_1} \langle\alpha'| V^*(t) |\alpha^0\rangle\, dt \right|^2$$

If for a certain transition $\langle\alpha'| V^* |\alpha^0\rangle$ is especially small or vanishing, the first approximation given above is inadequate. The next order approximation is given by

$$P(\alpha^0\alpha') = \frac{1}{\hbar^2}\Big|\int_{t_0}^{t'}\langle\alpha'|V^*(t)|\alpha^0\rangle\, dt$$
$$-\frac{i}{\hbar}\sum_{\alpha''}\int_{t_0}^{t'}\langle\alpha'|V^*(t_1)|\alpha''\rangle\, dt_1\int_{t_0}^{t_1}\langle\alpha''|V^*(t)|\alpha^0\rangle\, dt\Big|^2.$$

In the summation with respect to α'' we may omit the terms with $\alpha''=\alpha'$ and $\alpha''=\alpha^0$, since the matrix element corresponding to these terms are small compared with the others, or vanishing.

The expression for the transition probability in the second approximation lends itself to immediate physical interpretation. The first term in the expression for the amplitude corresponds to direct transitions between the states $|\alpha^0\rangle$ and $|\alpha'\rangle$, while the second term refers to transitions via an intermediate state $|\alpha''\rangle$, i.e. a transition first from $\alpha^0\rangle$ to $\alpha''\rangle$, followed by one form $|\alpha''\rangle$ to $\alpha'\rangle$. Since the amplitudes of these two-step transitions and of the direct transitions are added together (instead of their square moduli) the various transition processes interfere with each other.

The preceding results are all valid irrespective of whether V depends on t or not. If however V is independent of time, the integrals can be evaluated explicitly. (Cf. the remarks on the time-dependence of the Hamiltonian in the last lecture to see the importance of this case).

If V is independent of time (Schrödinger picture), so that $\langle\alpha'|V|\alpha^0\rangle$ is also time-independent, $\langle\alpha'|V^*|\alpha^0\rangle$ will have a time-dependence of the form $e^{i(E'-E^\circ)(t-t_0)/\hbar}\langle\alpha'|V|\alpha^\circ\rangle$.

If, for convenience we take $t_0=0$, the integration of the matrix element can be carried out to obtain

$$\int_0^{t'_1}\langle\alpha'|V^*|\alpha^\circ\rangle\,dt=\langle\alpha'|V|\alpha^\circ\rangle\,\frac{1-e^{i(E'-E^\circ)t'/\hbar}}{i(E'-E^\circ)/\hbar}.$$

Substituting this expression leads to

$$P(\alpha^\circ\alpha')=|\langle\alpha'|V|\alpha^\circ\rangle|^2\,\frac{2\{1-\cos((E'-E^\circ)t/\hbar)\}}{(E'-E^\circ)^2}.$$

If we plot P (α°, α') as a function of t' we get a curve as shown in Fig. 1 (a), which fluctuates with a frequency depending on $E'-E^\circ$. If we choose a different α' so that $E'-E_0$ is smaller, the frequency of the curve would be decreased. In the limiting case of $E'-E^\circ$ vanishing, the second factor of $P(\alpha^\circ\alpha')$ becomes simply $t'^2/\hbar^2$.

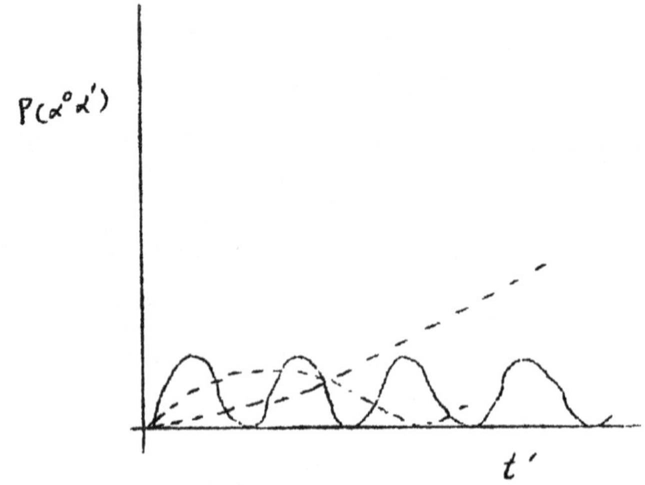

Fig. 1 a

Of course, this formula will break down for too large a value of t'.

It is gratifying to see that large changes in energy are less probable and vice versa.

This is a consequence of conservation of energy, which means conservation of $E+V$, and leads to approximate conservation of E.

For $E' = E^0$, we have the transition probability increasing as the square of the time interval (rather than being proportional to it). This result is at first sight difficult to understand and needs further discussion.

Consider the formula for the case of a continuous range of eigenvalues for the final energy E' and consider this final energy to vary, keeping close to the initial energy E^0. In this variation procedure we may neglect the variation of the first factor $|\langle\alpha'|V|\alpha^0\rangle|^2$. The probability of transition from $|\alpha^0\rangle$ to a final state of energy E' then varies with E' as shown in the figure. The quantity of physical importance is now the total transition probability to any final state with energy close to E^0, i.e. $\int P(\alpha^0\alpha')\, dE'$, and is given by the area under the curve in the figure.

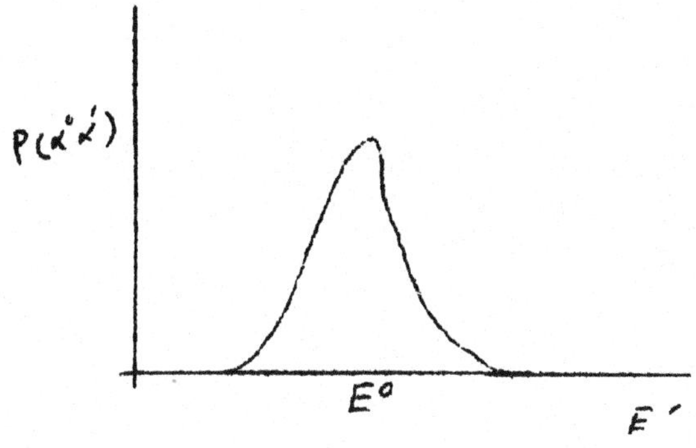

Fig. 1 b

As t' increases the peak value increases as t'^2 but the curve becomes * narrower; so that the area under the curve becomes proportional to t'. Thus the probability of a transition per unit time interval when integrated over all final energies in the neighbourhood of the initial energy is a constant.

---

* The algebra involved in the integration is as follows:

$$\int_0^\infty \left\{1 - \cos \frac{E' - E^\circ}{\hbar} t' \right\} / (E' - E^\circ)^2 \, dE'$$

$$= 2 \int_0^\infty \sin^2 \frac{(E' - E^\circ) t'}{2\hbar} / (E' - E^\circ)^2 \, dE'$$

$$= 2 \left(\frac{t'}{2\hbar}\right) \int_0^\infty \frac{\sin^2 x}{x^2} \, dx = \frac{\pi t'}{2\hbar}.$$

Lecture 13

In the last lecture we derived a formula for the transition probability of a system described by the Hamiltonian

$$H = E + V \qquad (1)$$

E being the proper energy of two things and V the small interaction between them. The state of the system was characterised by means of a complete set of commuting observables α which are constants of the motion for the unperturbed system whose Hamiltonian is E. The α's must therefore commute with E.

When the final state is in a continuous range then the quantity of physical interest is the total probability of transition into any state with energy close to E_0, i. e.

$$\int P(\alpha_0\, \alpha_0')\, dE' . \qquad (2)$$

To calculate this, it is convenient to choose for our α's the operator E itself and others which commute with E and are independent of it, which we call β. For example, in a scattering problem E would be the energy of the incident particle and β would be the variables that fix the direction of the incident particle.

We may now write

$$P(\alpha_0, \alpha') = P(\alpha_0, E'\beta') ,$$

and (2) becomes

$$\int P(\alpha^0, E', \beta') dE' \qquad (3)$$

In the case when H does not involve the time this integration can be performed and we have in the first approximation

$$\frac{2\pi}{\hbar} |\langle E_0 \beta' | V | \alpha^0 \rangle|^2 \, t \qquad (4)$$

It is important to note that E' has been replaced by E_0 on the left hand side of the matrix element. We now have a transition probability per unit time,

$$\frac{2\pi}{\hbar} |\langle E_0 \beta' | V | \alpha^0 \rangle|^2 . \qquad (5)$$

In the last lecture we saw that the curve showing the variation of $P(\alpha^0; E', \beta')$ on E' has a maximum at E_0. This maximum becomes sharper with the progress of time; the area under the curve increasing proportionally to the time, while the height increases proportionally to the square of the time. One can understand this behaviour from a physical point of view by considering again the example of a scattering process. When a particle of definite energy is scattered by a scatterer, there arise at the latter a train of scattered waves. The length of the train increases with the lapse of time. The energy of the scattered particle will be defined with lesser certainty at earlier times, when the scattered waves form only a short train of waves which cannot be sharply monochromatic than at later times when the scattered waves reach out farther and become more and more sharply monochromatic.

To remove the asymmetry between the initial and final states in the formula (5), due to the appearance of E_0 instead of E' on the left, we may replace the expression for $P(\alpha^0, \alpha')$ by

$$\frac{2\pi}{\hbar} |\langle E'\beta' |V| \alpha^0\rangle|^2 \, \delta(E'-E_0)$$

$$= \frac{2\pi}{\hbar} |\langle \alpha' |V| \alpha^0\rangle|^2 \, \delta(E'-E_0). \qquad (6)$$

This formula is an approximation inasmuch as it does not give correctly the dependence of $P(\alpha^0, \alpha')$ on E', but gives an infinite maximum at $E' = E_0$. But it is valid when integration over E' is performed.

The formula for transition probability to the next approximation is

$$\frac{2\pi}{\hbar} \left| \langle\alpha' |V| \alpha^0\rangle - \sum_{\substack{\alpha'' \neq \alpha^0 \\ \alpha'' \neq \alpha'}} \frac{\langle\alpha' |V| \alpha''\rangle \langle\alpha'' |V| \alpha^0\rangle}{E'' - E_0} \right|^2$$

$$\times \, \delta(E' - E_0). \qquad (7)$$

It can be interpreted in terms of transitions to intermediate states The denominator shows that those intermediate states whose energy is close to E_0 are the more important ones.

------x------

Let us now consider a system of n identical particles; let q_i be a complete set of commuting observables describing the ith particle, all q_i being of the same nature. Then we can construct a wave function

$$\psi(q_1, q_2, \dots q_n)$$

for the system. The Hamiltonian of the system must be symmetric, for otherwise the interchange of similar particles would change the energy. If we set up the Schrödinger picture of the motion, then the variation of ψ is determined by this symmetric Hamiltonian, so ψ must maintain whatever symmetry property it might have started with. We shall confine ourselves to only two possibilities, viz.,

(1) ψ is antisymmetric,

(2) it is symmetric.

It may be that for a certain kind of particle only wave functions with one of these symmetries occur in nature.

Let each of the n particles be in one of n distinct states defined by the wave functions

$$f_1(q),\ f_2(q),\ \ldots,\ f_n(q).$$

Then

$$\psi(q_1, \ldots, q_n) = f_1(q_1) \cdots f_n(q_n)$$

is a wave function for the whole system. To make it antisymmetric we permute the q's and put a plus or minus sign before the result according as the permutation is even or odd, respectively. Adding all the $n!$ such terms we obtain an antisymmetric function which may be expressed as the determinant

$$\psi_{anti} = \begin{vmatrix} f_1(q_1) & f_1(q_2) & \cdots & f_1(q_n) \\ f_2(q_1) & f_2(q_2) & \cdots & f_n(q_n) \\ & \cdots & & \\ f_n(q_1) & f_n(q_2) & \cdots & f_n(q_n) \end{vmatrix} . \qquad (10)$$

From (10) we see that if any two particles are in the same state, say $f_1 = f_2$, then the $\psi_{\text{anti}} \equiv 0$. Thus for particles which are described by antisymmetric wavefunctions we have the rule that no two of them can be in the same state. Thus Pauli's exclusion principle for electrons is derivable from the principles of quantum mechanics and the restriction of antisymmetric wavefunctions to describe the electrons. All particles of this kind obey Fermi statistics and are called fermions.

There are other kinds of particle for which only symmetrical wave functions are admissible. They obey Bose statistics and are called bosons.

All particles in nature are either fermions or bosons. There does not seem to be any theoretical reason for this, though one could mathematically construct wavefunctions with symmetry properties of a more complicated type.

There is a general rule that particles of half-odd-integral spins are fermions and those of integral spins are bosons. This rule is self consistent. For consider two particles of spin $\frac{1}{2}\hbar$ each. The combined system consisting of both of them bound together will have integral spin, and a wavefunction describing an assembly of them will be symmetric.

The rule has not been strictly proved, although it is much easier to construct relativistic theories of particles conforming to it than disobeying it. It has to be taken as an empirical result.

We shall now consider the situation when the number of particles is not fixed so that we have the possibility of creation and absorption of particles. For this purpose we shall need the concepts of addition and multiplication of vector spaces which we now introduce. Let V and W

be n and m dimensional vector spaces whose vectors are defined by means of coordinates $(x_1, x_2, \ldots, x_n)$ and $(y_1, y_2, \ldots, y_m)$ respectively. The sum $V + W$ is defined as the vector space spanned by all vectors of the form $(x_1, x_2, \ldots x_n, y_1, y_2, \ldots, y_m)$ This space is $n+m$ dimensional. The product $V \times W$ is defined to be the vector space spanned by all vectors

$$(x_r\, y_s) = (x_1 y_1, x_2 y_2, \ldots x_1 y_m, x_2 y_1, x_2 y_2 \ldots x_2 y_m, \ldots x_n y_1, x_n y_2, \ldots, x_n y_m).$$

This space is nm dimensional.

The wavefunction $f(q)$ can be regarded as a vector in the ket vector space whose coordinates are the numbers $f(q)$ for various points in the domain of q, q corresponding to the index r above. When we consider a system of n particles we have to consider wavefunctions like

$$f_1(q_1) \ldots f_n(q_n)$$

This will be a vector in the product space of the ket vector spaces of the individual particles. It will of course not be a general vector of this space.

Now let the number n of particles vary. We set up a ket vector space for each n. Adding these vector spaces we obtain a bigger space which describes a system with a variable number of particles.

Let the state with no particle at all be represented by the ket $|\cdot\rangle$ The space describing just one particle is spanned by kets $|f_r\rangle$, that with two particles by kets $S\left[|f_r\rangle\, |f_s\rangle\right]$ and so on. Here S

is an operator which produces suitable symmetry properties in the wave-functions. Adding all these spaces we obtain a mathematical formalism for dealing with a system with a variable number of particles. One can introduce emission and absorption operators, which vary the number of particles present.

In concluding this last lecture of this term we shall briefly summarise the salient features. The principle of superposition of states leads us to represent them by vectors in a generalized Hilbert space. The dynamical variables are represented by linear operators which operate on these vectors. They are governed by an anticommutative algebra. This non-commutativity leads to uncertainty relationships when we try to give numerical values to two dynamical variables at the same time, unless they commute. We then have general principles connecting rotation operators with the angular momentum of a system and also translational operators with the linear momentum. We introduced the laws of motion in the Schrodinger and the Heisenberg forms, which are mathematically equivalent and connected by a unitary transformation. Finally there are the restrictions which limit wave-functions for systems of identical particles to be symmetric or antisymmetric.

The only question which remains is that of the choice of the Hamiltonian for a given problem. For electrons interacting with electromagnetic field we have a quite good Hamiltonian which yields very accurate results. But for meson theories we do not have accurate Hamiltonians.

In the work of finding the Hamiltonian we are guided by classical mechanics. When dealing with spin we cannot rely on classical mechanics, for the classical mechanical concept of spin is inadequate. But we can invoke the principle of relativity for the purpose.

It is necessary that an accurate theory should conform to the principle of relativity. This imposes severe restrictions in our search for a Hamiltonian. It requires us usually to have interactions only through the intermediary of fields. One can have relativistic interactions at a distance in special cases, but these cases need a complicated transformation theory to show their relativistic invariance.

The introduction of fields introduces complications because now the number of variables becomes infinite. We shall start a study of them in the next term.

Lecture 14.

The laws of quantum mechanics must be so formulated that they satisfy the principle of special relativity, i.e. they must be the same for two observers in uniform relative motion. This requirement becomes important when dealing with particles whose velocities are near that of light. It is not important to make the laws of quantum mechanics obey the principle of general relativity, since the latter is needed only to explain gravitational effects, which are extremely small for atomic systems.

We shall first fit in the classical mechanics in Hamiltonian form with the special theory of relativity. For this purpose it is most convenient to treat the space and time coordinates on the same footing and to regard the physical world as a four dimensional space, with coordinates $x_1, x_2, x_3, x_0 (= t)$, the velocity of light being taken to be unity.

Let $q_r, p_r, r = 1, 2, \ldots, n$ be the coordinates and the conjugate momenta of a dynamical system, and $\xi = \xi(q, p)$ be a dynamical variable. Then the equation of motion for ξ is

$$\frac{d\xi}{dt} = [\xi, H] \tag{1}$$

where $[A, B]$ stands for the Poisson Bracket. This equation is not invariant under Lorenz transformations since it treats time as a special variable.

Let us now regard t as a another coordinate and introduce the variable

$$w = -H \tag{2}$$

as its conjugate momentum. Consider a general dynamical variable to be

$$\eta = \eta(q, p, t, w). \tag{3}$$

Then the P.B. will be defined by

$$[\eta_1, \eta_2] = \sum \left(\frac{\partial \eta_1}{\partial q_r} \frac{\partial \eta_2}{\partial p_r} - \frac{\partial \eta_1}{\partial p_r} \frac{\partial \eta_2}{\partial q_r} \right) + \frac{\partial \eta_1}{\partial t} \frac{\partial \eta_2}{\partial w} - \frac{\partial \eta_1}{\partial w} \frac{\partial \eta_2}{\partial t}. \tag{4}$$

Define

$$\mathcal{H} = H + W = 0. \tag{5}$$

Then

$$\begin{aligned} \frac{d\eta}{dt} &= \sum_r \left(\frac{\partial \eta}{\partial q_r} \dot{q}_r + \frac{\partial \eta}{\partial p_r} \dot{p}_r \right) + \frac{\partial \eta}{\partial t} + \frac{\partial \eta}{\partial w} \dot{w} \\ &= \sum_r \left(\frac{\partial \eta}{\partial q_r} \frac{\partial H}{\partial p_r} - \frac{\partial \eta}{\partial p_r} \frac{\partial H}{\partial q_r} \right) + \frac{\partial \eta}{\partial t} + \frac{\partial \eta}{\partial w} \dot{w} \\ &= \sum_r \left(\frac{\partial \eta}{\partial q_r} \frac{\partial \mathcal{H}}{\partial p_r} - \frac{\partial \eta}{\partial p_r} \frac{\partial \mathcal{H}}{\partial q_r} \right) + \frac{\partial \eta}{\partial t} \frac{\partial \mathcal{H}}{\partial w} - \frac{\partial \eta}{\partial w} \frac{\partial \mathcal{H}}{\partial t} \\ &= [\eta, \mathcal{H}]. \end{aligned} \tag{6}$$

We have here used the relation

$$\frac{\partial \mathcal{H}}{\partial t} = \frac{\partial H}{\partial t} = \frac{dH}{dt} = -\frac{dW}{dt}. \tag{7}$$

Now introduce a general independent variable τ,

$$\tau = \tau(q, p, t, w)$$

such that $\frac{d\tau}{dt} \neq 0$. Then

$$\begin{aligned} \frac{d\eta}{d\tau} &= \frac{d\eta}{dt} \cdot \frac{dt}{d\tau} = [\eta, \mathcal{H}] \frac{dt}{d\tau} + \mathcal{H} \left[\eta, \frac{dt}{d\tau}\right] \\ &= [\eta, \mathcal{H}^*], \end{aligned} \tag{8}$$

where

$$\mathcal{H}^* = \mathcal{H} \frac{dt}{d\tau}, \tag{9}$$

We now have complete symmetry between t and the q's.

The equation $\mathcal{H}^* = 0$ should only be used after the evaluation of the Poisson bracket. Such equations are called weak equations and will be written

$$\mathcal{H}^* \approx 0 \tag{10}$$

$\mathcal{H}^*$ can be multiplied by any factor, causing a change in τ. Conversely, fixing τ will give a definite factor to multiply $\mathcal{H}^*$.

To go over into quantum mechanics, q_r, p_r, t, w are made operators satisfying commutation relations corresponding to the P.B. relations. Wave functions must satisfy the equation

$$\begin{aligned} &\mathcal{H}\psi = 0, \\ \text{or } &(H + W)\psi = 0, \qquad W = -i\hbar \frac{\partial}{\partial t}, \end{aligned} \tag{11}$$

which is the Schrödinger equation. More generally,

$$\mathcal{H}^* \psi = 0 \tag{12}$$

provided that $\mathcal{H}^* = \lambda \mathcal{H}$, with λ on the left. The weak equation $\mathcal{H}^* \approx 0$ is to be interpreted as $\mathcal{H}^* \psi = 0$ in quantum mechanics. Ordinary equations or strong equations are equations between operators while weak equations are conditions on wave functions.

We shall now consider a single particle whose coordinates are x_0, x_1, x_2, x_3, with $-p_0, p_1, p_2, p_3$ the momenta conjugate to them. There may in addition be internal variables if the particle has structure. An arbitrary point in space time will be denoted by x_μ ($\mu = 0, 1, 2, 3$).

Layer indices (A_μ) denote contravariant vectors i.e., one transforming like x_μ or z_μ, while upper indices (A^μ) denote co-variant vectors transforming like $\frac{dV}{dx_\mu}$. For the space-time of special relativity we have:

$$A_0 = A^0, \quad A_r = - A^r, \quad r = 1, 2, 3, \tag{13}$$

so that the metric tensor $g_{\mu\nu}$ is

$$\left.\begin{array}{ll} g_{\mu\nu} = 0 & \text{for} \quad \mu \neq \nu \\ g_{00} = 1, & g_{rr} = -1. \end{array}\right\} \tag{14}$$

The commutation relations between p_μ and z_ν are

$$[p_\mu, z_\nu] = g_{\mu\nu}, \tag{15}$$

i.e.,

$$[p_0, z_0] = 1, \quad [p_r, z_r] = -1.$$

Thus p_0 is the negative of the conjugate momentum to z_0, and is therefore the energy.

For a free particle the Hamiltonian is

$$p_0^2 - p_1^2 - p_2^2 - p_3^2 - m^2 \approx 0, \tag{16}$$

where m is the mass of the particle. Take

$$\mathcal{H} = -\frac{1}{2m}(p_\mu p^\mu - m^2) \approx 0. \tag{17}$$

This choice of $\mathcal{H}$ makes $\tau = \pm s$, the proper time, since

$$\frac{dz_\mu}{d\tau} = [z_\mu, \mathcal{H}] = -\frac{1}{m} p_\nu [z_\mu, p^\nu] = \frac{1}{m} p_\mu, \tag{18}$$

whence

$$\frac{dz_\mu}{d\tau} \frac{dz^\mu}{d\tau} = 1, \tag{19}$$

so that

$$\frac{dz_\mu}{d\tau} = \pm \frac{dz_\mu}{ds} \quad \text{, and hence} \quad \tau = \pm s .$$

If p_0 is +ve, then $\frac{dz_0}{ds}$ is +ve and $\tau = s$. If p_0 is negative, $\tau = -s$.

To go over to quantum theory we replace p_μ by

$$p_\mu = -i \hbar \frac{\partial}{\partial x^\mu} , \qquad (20)$$

whence we obtain the wave equation

$$\left\{ \hbar^2 \frac{\partial}{\partial x_\mu} \frac{\partial}{\partial x^\mu} + m^2 \right\} \psi = 0 . \qquad (21)$$

The Schrödinger egation that we set up last term was linear in $\frac{d}{dt}$. There it was assumed that the position of the particle was an observable. Equation (21) describes a particle whose position is not an observable. One cannot introduce into this theory the probability of the particle being found in a specified small volume.

The ψ of equation (21) is a scalar quantity and describes a particle with no spin.

We can obtain a relativistic equation for a particle which is linear in $\frac{d}{dt}$ if internal variables are introduced. We assume it to be

$$(p_0 + \alpha_1 p_1 + \alpha_2 p_2 + \alpha_3 p_3 + m \alpha_m) \psi = 0 . \qquad (22)$$

The α's are new dynamical variables which commute with z, p. We require that (21) be a consequence of (22). This is so if α's satisfy the commutation relations

$$\alpha_1^2 = \alpha_2^2 = \alpha_3^2 = \alpha_m^2 = 1$$

$$\alpha\text{'s anti-commute ,} \qquad (23)$$

as can be verified by multiplying (22) by $(p_0 - \alpha_1 p_1 - \alpha_2 p_2 - \alpha_3 p_3 - \alpha_m m$ on the left. This equation is relativistic and describes a particle of spin $\frac{1}{2}\hbar$, as will be shown in the next lecture. The α's can be represented by 4x4 matrices, and ψ is a four component quantity.

We can set up other types of wave equation which are relativistic. Let the wave function be a vector quantity ψ_μ, 0,1,2,3. We put

$$F_{\mu\nu} = \frac{\partial \psi_\mu}{\partial x^\nu} - \frac{\partial \psi_\nu}{\partial x^\mu},$$

$$\frac{\partial}{\partial x_\mu} F_{\mu\nu} = \frac{m^2}{\hbar^2} \psi_\nu . \qquad (24\ a,b)$$

From (24 b) we obtain

$$\frac{\partial \psi_\nu}{\partial x_\nu} = 0 \qquad \text{for } m \neq 0, \qquad (25)$$

and then (24a) gives

$$-\frac{\partial^2}{\partial x_\mu \partial x^\mu} \psi_\nu = \frac{m^2}{\hbar^2} \psi_\nu . \qquad (26)$$

This equation describes a particle of spin $\hbar$ since the independent components of the wave function ψ_1, ψ_2, ψ_3 transform as an ordinary vector for rotations.

For the case $m = 0$ the relation $\frac{\partial \psi_\mu}{\partial x_\mu} = 0$ is not derivable from (24) and the equation has to be treated in a different manner. It describes the photon which has spin $\hbar$.

Lecture 15.

We have seen that the relativistic wave equation of a single particle (for which the position is an observable) may be written in the form

(1) $$(p_0 + \alpha^r p_r + \alpha_m m)\psi = 0, \quad r = 1, 2, 3 \qquad 1,2,3$$

where the α 's satisfy the relations

(2) $$(\alpha^r)^2 = 1, \quad (\alpha_m)^2 = 1$$

α 's anti commute,

By multiplying this equation on the left by α_m we may write this in the form

(3) $$(\gamma^\mu p_\mu + m)\psi = 0 \qquad , \mu = 0, 1, 2, 3$$

where the γ^μ are defined by the equations

$$\gamma^0 = \alpha_m, \quad \gamma^r = \alpha_m \alpha^r, \qquad r = 1,2,3.$$

It is easily verified that these γ's satisfy the relations

$$(\gamma^0)^2 = 1, \quad (\gamma^r)^2 = \alpha_m \alpha^r \alpha_m \alpha^r = -1,$$

γ 's anti commute .

These relations may be summarised into a single equation:

(4) $$\gamma^\mu \gamma^\nu + \gamma^\nu \gamma^\mu = 2g^{\mu\nu}.$$

The relativistic wave equation can thus be expressed by the equations (3) and (4) instead of the equations (1) and (2). Now equations (3) and (4) are expressed correctly in tensor notation (with suffixes on both sides balancing). This had had the effect that the Lorentz invariance of the wave-equation is often incorrectly discussed. It is argued that the p 's and γ 's transform like 4-vectors, and so preserve the invariance of the equations (3) and (4), on the assumption that ψ is an invariant.

This is procedure is incorrect, since it does not take into account the Hermitian character of the γ 's. To see this, we make use of the result that it is possible to construct the γ 's from two independent sets of Pauli matrices $\sigma_1, \sigma_2, \sigma_3$ and ρ_1, ρ_2, ρ_3. Take

$$\alpha_1 = \rho_1 \sigma_1$$

$$\alpha_2 = \rho_1 \sigma_2$$

$$\alpha_3 = \rho_1 \sigma_3$$

$$\alpha_m = \rho_3$$

Then these α 's satisfy all the requirements(2). Since the Pauli matrices are Hermitian, this construction shows that the α 's are also Hermitian. Hence γ^0 is Hermitian:

$$\bar{\gamma}^0 = \gamma^0$$

On the other hand

$$\bar{\gamma}^r = \bar{\alpha}^r \bar{\alpha}^m = \alpha^r \alpha^m = -\gamma^r \qquad r = 1,2,3.$$

Since the α 's anti commute. Hence we see that γ^0 is Hermitian and γ^r are anti-Hermitian and consequently the γ 's cannot transform like the components of a 4-vector. This discrepancy in the Hermitian properties of the γ 's is independent of the representation employed, since

$$(\gamma^r)^2 = -1$$

and hence γ^r cannot be Hermitian.

The Lorentz invariance of the wave-equation is to be proved by other methods. We can, for example, consider typical Lorentz transformations and explicitly show that the equation is invariant under each of these.* But this method is tedious, and a new method employing infinitesimal transformations is presented below:**

Consider any general infinitesimal transformation

$$p_\mu \rightarrow p_\mu^* = p_\mu + \varepsilon\, m_\mu{}^\nu p_\nu$$

where ε is infinitesimal. If this is to be a Lorentz transformation we should have

$$p_\mu p^\mu = p_\mu^* p^{*\mu} = p_\mu p^\mu + 2\varepsilon\, m_\mu{}^\nu p_\nu p^\mu$$

retaining terms to first order only. This requires

$$2\varepsilon\, m_\mu{}^\nu p_\nu p^\mu = 2\varepsilon\, m_{\mu\nu} p^\mu p^\nu = 0$$

so that we should have

$$m_{\mu\nu} = - m_{\nu\mu}.$$

Define

$$\alpha^0 = 1$$

Then, the wave equation (1) may be written

$$(\alpha^\mu p_\mu + \alpha_m m)\psi = 0 .$$

We define the lowering of the suffixes of the α's in the usual way,

$$\alpha_0 = \alpha^0 , \qquad \alpha_r = -\alpha^r$$

---

* This is done in § 68 of Principles of Quantum Mechanics.

** We discard the misleading γ-notation and prefer to work with α's only.

Then one can easily verify the following relation to be correct:

$$\alpha_\mu \alpha^\nu + \alpha_\nu \alpha^\mu = 2 g_{\mu\nu} = 2 g^{\mu\nu} .$$

Here we have an equation with suffixes unbalanced on either side and it cannot therefore be a tensor equation. It is to be proved by considering separately the cases $\mu = \nu = 0$; $\mu = 0, \nu = r$; $\mu = r, \nu = s$. Other similar correct equations with unbalanced suffixes are

$$\alpha^\mu \alpha_\nu + \alpha^\nu \alpha_\mu = 2 g_{\mu\nu} = 2 g^{\mu\nu}$$

$$\alpha_\mu \alpha_m = \alpha_m \alpha^\mu$$

$$\alpha^\mu \alpha_m = \alpha_m \alpha_\mu$$

We now apply a general infinitesimal Lorentz transformation. The α's are fixed matrices while the p_μ are changed to p_μ^*. We have

$$\{ \alpha^\mu (p_\mu^* - \epsilon\, m_{\mu\nu} p_\nu^*) + \alpha_m m \} \psi = 0 .$$

Multiply this equation by the factor

$$(1 + \tfrac{1}{4} \epsilon\, m_{\rho\sigma} \alpha^\rho \alpha_\sigma)$$

from the left:

$$(1 + \tfrac{1}{4} \epsilon\, m_{\rho\sigma} \alpha^\rho \alpha_\sigma) \{ \alpha^\mu (p_\mu^* - \epsilon\, m_\mu{}^\nu p_\nu^*) + \alpha_m m \} \psi = 0.$$

Now,

$$\tfrac{1}{4} \epsilon\, m_{\rho\sigma} \alpha^\rho \alpha_\sigma \alpha^\mu p_\mu^* = \tfrac{1}{4} \epsilon\, m_{\rho\sigma} \alpha^\rho (2 g^{\sigma\mu} - \alpha_\mu \alpha^\sigma) p_\mu^*$$

$$= \tfrac{1}{4} \epsilon\, m_{\rho\sigma} \{ 2 \alpha^\rho g^{\sigma\mu} - (2 g^{\rho\mu} - \alpha^\mu \alpha_\rho) \alpha^\sigma \} p_\mu^*$$

$$= \epsilon\, m_\rho{}^\mu \alpha^\rho + \tfrac{1}{4} \epsilon\, m_{\rho\sigma} \alpha^\mu \alpha_\rho \alpha^\sigma p_\mu^* ,$$

where use has been made of the antisymmetry of $m_{\mu\nu}$. The first term of the last member of the chain of equalities cancels a similar term in the complete product. Further, we have

$$\alpha^{\rho}\alpha_{\sigma}\alpha_m = \alpha_m \alpha_{\rho}\alpha^{\sigma}.$$

Hence, we get

$$\{\alpha^{\mu} p_{\mu}^{*} + \tfrac{1}{4}\varepsilon\, m_{\rho\sigma}\alpha^{\mu}\alpha_{\rho}\alpha^{\sigma} p_{\mu}^{*} + \alpha_m m(1+\tfrac{1}{4}\varepsilon\, m_{\rho\sigma}\alpha_{\rho}\alpha^{\sigma})\}\psi = 0$$

or

$$(\alpha^{\mu} p_{\mu}^{*} + \alpha_m m)(1+\tfrac{1}{4}\varepsilon\, m_{\rho\sigma}\alpha_{\rho}\alpha^{\sigma})\psi = 0.$$

which can be written in the form

$$(\alpha^{\mu} p_{\mu}^{*} + \alpha_m m)\psi^{*} = 0$$

with

$$\psi^{*} = (1+\tfrac{1}{4}\varepsilon\, m_{\rho\sigma}\alpha_{\rho}\alpha^{\sigma})\psi .$$

This form is identical with the equation we started with; and hence the Lorentz invariance is proved. It is to be noted that in this transformation, the α 's are unchanged, p_{μ} transform like the components of a 4-vector while ψ transforms in a particular manner. We will see presently that this is the transformation law appropriate to the wave-function of spin $\frac{1}{2}$ particle.

In the above proof of relativistic invariance, we have made extensive use of equations with unbalanced suffixes. Such equations are necessary for dealing with spin $\frac{1}{2}$ particles as the usual tensor analysis is inadequate for this purpose. (One would recall the result that a rotation through 2π introduced a factor -1 into the wave-function of a spin $\frac{1}{2}$ particle.)

To verify that the above equation really describes a particle of spin $\frac{1}{2}$ we make use of the general relationship between angular momentum and rotation operators. Apply a small rotation through an angle ϵ about the x_1-axis. Then goes over into

$$\psi^* = (1 + \epsilon \gamma_1)\psi$$

and
$$m_1 = i\hbar\gamma_1$$

(as derived earlier in this course of lectures). This rotation is a particular case of a Lorentz transformation, with

$$m_{23} = -m_{32} = 1$$

and other components of $m_{\mu\nu}$ zero. Putting in these values for $m_{\mu\nu}$, we find that the transformed value of ψ is

$$\psi^* = (1 - \tfrac{1}{2}\epsilon\,\alpha_2\alpha_3)\psi$$

Since there are only 2 non-vanishing terms in the sum $m_{\mu\nu}\alpha_\mu\alpha^\nu$. Comparing the two values of ψ^*, we obtain,

$$\gamma_1 = -\tfrac{1}{2}\alpha_2\alpha_3 \qquad m_1 = -\tfrac{1}{2} i\hbar\,\alpha_2\alpha_3$$

Now,
$$(i\,\alpha_2\alpha_3)^2 = 1$$

and hence $i\alpha_2\alpha_3$ has the eigenvalues ± 1. Hence m_1 has the eigenvalues $\pm\frac{1}{2}\hbar$. Thus we have verified that the wavefunction ψ describes a particle of spin half a quantum.

So far we have concerned ourselves with the basic but trivial case of a free particle. Consider now an electron with charge $-e$ in a given electromagnetic field. The Hamiltonian equation is the weak equation

$$\mathcal{H} = -\frac{1}{2m}\{(p_\mu + eA_\mu)(p^\mu + eA^\mu) - m^2\} \approx 0$$

where

$$A_\mu = A_\mu(x)$$

is the electromagnetic potential at the point z where the electron is situated. To verify that this is the correct Hamiltonian, we deduce the equations of motion:

$$\frac{dz_\mu}{d\tau} = [z_\mu, \mathcal{H}] = -\frac{1}{m}[z_\mu, p_\nu + eA_\nu](p^\nu + eA^\nu)$$

$$= \frac{1}{m}(p_\mu + eA_\mu)$$

or

$$p_\mu + eA_\mu = m\frac{dz_\mu}{d\tau}.$$

The weak equation

$$\mathcal{H} \approx 0$$

gives

$$\frac{dz_\mu}{d\tau} \cdot \frac{dz^\mu}{d\tau} = 1$$

showing that

$$\tau = \pm s$$

where s is the proper time. Confining ourselves to the case $\tau = s$, which occurs when the energy is positive, we have

$$\frac{dp_\mu}{ds} = [p_\mu, \mathcal{H}] = -\frac{1}{m}[p_\mu, p_\nu + eA_\nu](p^\nu + eA^\nu)$$

$$= -\frac{dz^\nu}{ds} \cdot e\left(\frac{\partial A_\nu}{\partial x^\mu}\right)_z$$

where the suffix z denotes that the quantities are to be evaluated at the point z. Substituting for p_μ in terms of velocity in the form

$$p_\mu = m\frac{dz_\mu}{ds} - eA_\mu$$

we obtain,

$$m\frac{d^2 z_\mu}{ds^2} = e\frac{dz^\nu}{ds}\left(-\frac{\partial A_\nu}{\partial x^\mu} + \frac{\partial A_\mu}{\partial x^\nu}\right) = -e\frac{dz^\nu}{ds}F_{\mu\nu}$$

where

$$F_{\mu\nu} = \frac{\partial A_\nu}{\partial x_\mu} - \frac{\partial A_\mu}{\partial x_\nu}$$

is the 6-vector which describes the electromagnetic field:

$$F_{10} = \frac{\partial A_0}{\partial x^1} - \frac{\partial A_1}{\partial x^0} = -\frac{\partial A_0}{\partial x_1} - \frac{\partial A_1}{\partial x_0} = \mathcal{E}_1$$

$$F_{12} = \frac{\partial A_2}{\partial x_1} - \frac{\partial A_1}{\partial x_2} = -\frac{\partial A_2}{\partial x_1} + \frac{\partial A_1}{\partial x_2} = -\mathcal{H}_3$$

$\mathcal{E}$ and $\mathcal{H}$ denoting the electric and magnetic field strengths.

The quantisation of this theory would give an electron without spin. The generalisation of the theory for describing an electron with spin $\frac{1}{2}$ is achieved by adopting the Hamiltonian

$$\mathcal{H} = \alpha^\mu (p_\mu + e A_\mu) + \alpha_m \cdot m$$

and the wave-equation

$$\mathcal{H} \psi = 0 .$$

Lecture 16.

We wish to set up a general relativistic theory of several particles interacting with one another. In a relativistic theory one cannot have interaction at a distance, but only through the intermediary of a field. A field has to be treated as a dynamical system, for the field itself will change due to its interaction with the particles. The various components of a field at all points of space (at a given time) will be regarded as dynamical variables. The number of dynamical variables is thus three-fold infinite. Apart from this, a field is fundamentally the same as a dynamical system consisting of particles.

We want to have Poisson bracket relations between field quantities. Let $\xi(x_1, x_2, x_3)$ and $\eta(x_1', x_2', x_3')$ be two field variables. Since they are observables and since the measurement of one need not affect that of the other if the space points x and x' are distinct, they are compatible and will commute. We therefore have

$$[\xi(x_1 x_2 x_3), \eta(x_1' x_2' x_3')]$$

= something involving $\delta(x_1 - x_1')\delta(x_2 - x_2') \times \delta(x_3 - x_3')$

and its derivatives,

or, as we shall write for brevity,

$$[\xi, \eta'] \quad = \text{something involving } \delta_3(x - x')$$

and its derivatives.

We must now set up a Hamiltonian dynamics involving the field and particle variables. Since we wish to build a relativistic theory we must not confine ourselves to defining the variables at a given time and finding them at a later one, i.e., passing from surface 1 to surface 2

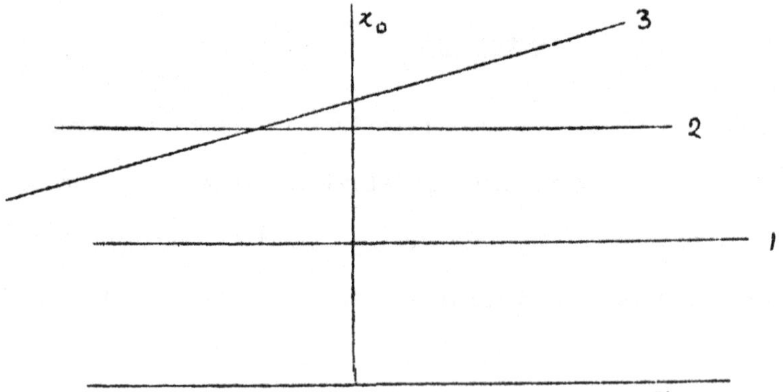

Fig. 1.

(cf. figure 1). For another observer surface 2 may appear as the surface 3, say. A relativistic theory must therefore allow one to pass from surface 1 to surface 3.

Although one can construct a relativistic theory using only flat surfaces, such a theory would not necessarily require the interaction between two particles at large distances to be small. This arises because a small change in the angle of a surface will cause large changes of time at large distances and so the effect of conditions at large distances becomes too prominent. In order to avoid these defects we shall consider our dynamical variables on curved space like surfaces

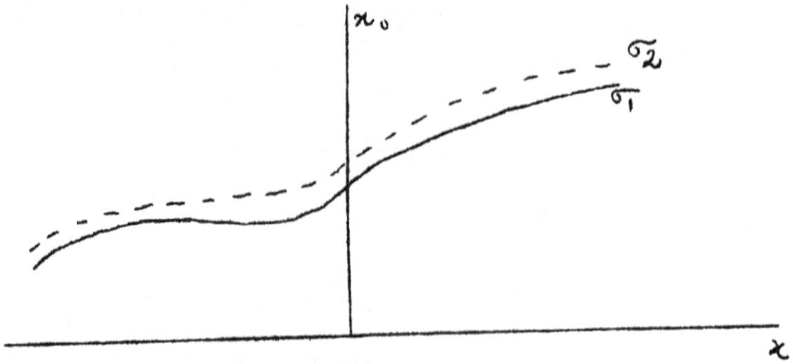

Fig. 2.

and set up equations of motion which will enable us to pass from one space like surface σ_1 to another σ_2.

To set up the equations of motion in a relativistic theory we need a generalisation of the usual Hamiltonian theory. We need several Hamiltonians

$$\mathcal{H}_1 \approx 0, \quad \mathcal{H}_2 \approx 0, \quad \ldots ,$$

with the corresponding independent variables $\tau_1, \tau_2, \ldots$ The equations of motion are then

$$\frac{d\xi}{d\tau_1} = [\xi, \mathcal{H}_1], \quad \frac{d\xi}{d\tau_2} = [\xi, \mathcal{H}_2]$$

The general motion will be described by

$$\frac{d\xi}{d\tau_{gen}} \approx \sum_r v_r [\xi, \mathcal{H}_r]$$

$$\approx \sum_r [\xi, v_r \mathcal{H}_r] = [\xi, \mathcal{H}_{gen}]$$

$$\mathcal{H}_{gen} = \sum v_r \mathcal{H}_r \approx 0 \qquad (1)$$

Here v_r are arbitrary coefficients. Because of the coefficients v_r there will be an arbitrariness in the solution of these equations.

Consider the case of flat surfaces. One needs in general four variables to change one flat surface into another. For each of them one should have a Hamiltonian. The general Hamiltonian then provides a fourfold arbitrariness in the motion of the surface.

For the case of curved surfaces we shall need an infinite number of

Hamiltonians since there are infinite number of ways of deforming such a surface.

When we have many Hamiltonians we have as many equations of motion. We have then to examine their consistency. The general consistency condition is

$$\frac{d^2 \xi}{d\tau_r d\tau_s} \approx \frac{d^2 \xi}{d\tau_s d\tau_r} \qquad \text{for all } \xi$$

whence

$$[[\xi, \mathcal{H}_s], \mathcal{H}_r] \approx [[\xi, \mathcal{H}_r], \mathcal{H}_s]$$

From the Jacobi identity we now obtain

$$[\xi, [\mathcal{H}_r, \mathcal{H}_s]] \approx 0 \qquad \text{for all } \xi$$

Hence

$$[\mathcal{H}_r, \mathcal{H}_s] = c\text{-number.}$$

A further requirement for consistency is that $\mathcal{H}_r$ shall always remain zero. Thus we must have

$$\frac{d\mathcal{H}_r}{d\tau_s} \approx 0 \qquad (2)$$

so that

$$[\mathcal{H}_r, \mathcal{H}_s] \approx 0.$$

This is a slightly stronger condition than we had previously.

Since the only things weakly equal to zero are the $\mathcal{H}_r$ and linear combination of them, this equation can be replaced by the strong equation

$$[\mathcal{H}_r, \mathcal{H}_s] = c_{rs}^{\ t}\, \mathcal{H}_t \, , \qquad (3)$$

$c_{rs}^{\ t}$ being any functions of the dynamical variables.

To go over to quantum mechanics one now regards the dynamical variables as operators satisfying commutation relations corresponding to their Poisson bracket relations. The various weak equations $\mathcal{H}_r \approx 0$ are replaced by corresponding Schrödinger equations $\mathcal{H}_r \psi \approx 0$.

The consistency conditions are now

$$\mathcal{H}_s \mathcal{H}_r \psi = 0$$
$$\mathcal{H}_r \mathcal{H}_s \psi = 0$$

whence

$$[\mathcal{H}_r, \mathcal{H}_s]\psi = 0 . \qquad (4)$$

$[\mathcal{H}_r, \mathcal{H}_s]$ must therefore be weakly equal to zero, and therefore strongly equal to $c_{rs}^{\ t}\, \mathcal{H}_t$, $c_{rs}^{\ t}$ being again functions of the dynamical variables, but now standing necessarily on the left of the $\mathcal{H}_t$

To set up a particular dynamical theory one must have explicitly the Hamiltonians $\mathcal{H}_r$. To find them may be a difficult matter. In classical mechanics one usually starts from a Lagrange function or an action integral from which a Hamiltonian and Hamiltonian equations of motion can be derived. By taking the action to be Lorentz invarient one can ensure that the theory is relativistic

One can get the general case of several Hamiltonians from a Lagrangian by proceeding in the following manner (P.A.M. Dirac, Canadian Journ.Maths.2 127). Let q_n , $n = 1, 2, \ldots N$ be the coordinates and $\dot{q}_n = \frac{dq_n}{dt}$

the corresponding velocities of a dynamical system. Let

$$L = L(q, \dot{q})$$

be a Lagrange function. The corresponding Euler equations are

$$\frac{d}{dt} \cdot \frac{\partial L}{\partial \dot{q}_n} = \frac{\partial L}{\partial q_n}, \quad n = 1, 2, \ldots, N. \qquad (6)$$

We now regard t as an extra coordinate $t = q_0$, with the velocity $\dot{q}_0 = 1$. We make L homogeneous and of the first degree in the velocities by means of $\dot{q}_0$ and call it $\tilde{L}$

$$\tilde{L} = \dot{q}_0 \, L(q, \frac{\dot{q}}{\dot{q}_0})$$

The action integral is then

$$I = \int \tilde{L} \, dt$$

Now introduce a new independent variable τ. We have

$$I = \int L^* \, d\tau$$

where

$$L^* = \frac{dq_0}{d\tau} L(q, \frac{\frac{dq}{d\tau}}{\frac{dq_0}{d\tau}}). \qquad (8)$$

This expression for the action treats the time on exactly the same footing as the other coordinates. The equations of motion obtained from varying this action will be (6), with τ substituted for t and $\frac{dq_n}{d\tau}$ for $\dot{q}_n$ and with n going from 0 to N.

Now define the momenta p_n by

$$p_n = \frac{\partial L^*}{\partial \dot{q}_n}, \quad n = 0, \ldots, N. \qquad (9)$$

where $\dot{q}_n = \frac{dq_n}{d\tau}$. The p_n will be homogeneous functions of the velocities of degree zero, i.e., will involve only ratios of the velocities. Since there are $n+1$ p's and only n ratios of velocities there will exist at least one relation between the p_r that does not involve the velocities. There may be more than one. We denote them by

$$\phi_m(p,q) \approx 0 \ , \quad m = 1, 2, \ldots, M. \qquad (10)$$

There may in addition be further relations between p and q arising from the equations of motion themselves, which we express as

$$\chi_k(p,q) \approx 0 \ , \quad k = 1, 2, \ldots, K. \qquad (11)$$

These ϕ and χ can be regarded as the Hamiltonians (as will be shown in the next lecture) provided the Poisson bracket of any two of them is weakly equal to zero. This is so in most cases of physical theories, the vector meson being an exception.

Lecture 17.

Starting with any given Lagrangian, it is possible to build up a system of generalised Hamiltonian dynamics. Firstly we make the Lagrangian homogeneous of the first degree in the velocities $\dot{q}$ by taking (as described in detail in the last lecture), the time t as an extra co-ordinate q_0. We shall now use the notation

$$\dot{q}_n = \frac{dq_n}{d\tau}, \quad n = 0, \ldots, N$$

where τ is an arbitrary independent variable and we have L homogeneous of the first degree in these velocities.

The momenta are defined in the usual manner by the relations

$$p_n = \frac{\partial L}{\partial \dot{q}_n}$$

Consider small independent variations of order ε in the q's, $\dot{q}$'s and p's. Then the equations defining p_n will be violated to order ε. We should now distinguish between two types of equations; equations which are not violated to the accuracy ε by a variation of order ε, called 'strong' equations (and written with the usual equality sign $=$); and equations which are violated to the accuracy ε by a variation of the order ε, called 'weak' equations (written with the equality sign $\approx$). Consistent with this notation we write the relation for the momenta as a weak equation

$$p_n \approx \frac{\partial L}{\partial \dot{q}_n}. \qquad (1)$$

In passing, we notice the following properties of weak equations:

If $A \approx 0$, then dA is not necessarily zero. But if $X = 0$, $dX = 0$

Also if $A \approx 0$,

$$d(A^2) = 2A\,dA = 0, \quad \text{so } A^2 = 0$$

More generally,

if $A \approx 0$ and $B \approx 0$, then $AB = 0$.

Suppose equations (1) lead to certain relations between the p's and q's; not involving velocities. These equations are weak equations, since they are consequences of weak equations.

Let the complete set of independent equations connecting the p's and q's (not involving the velocities) which follow from the defining equation of the p_n, be denoted by

$$\phi_m \approx 0, \quad m = 1, 2, \ldots, M$$

Since the Lagrangian is homogeneous of the first degree in the velocities, Euler's theorem on homogeneous functions gives

$$L = \dot{q}_n \frac{\partial L}{\partial \dot{q}_n} \approx \dot{q}_n p_n$$

Here the first equality is a strong equation since it is independent of the momenta. Define the Hamiltonian $\mathcal{H}$ by the usual relation:

$$\mathcal{H} = \dot{q}_n p_n - L \approx 0$$

The Hamiltonian so introduced is weakly equal to zero. Now make a general variation of the q's, $\dot{q}$'s and p's.

Then,

$$\delta \mathcal{H} = -\frac{\partial L}{\partial q_n}\delta q_n - \frac{\partial L}{\partial \dot{q}_n}\delta \dot{q}_n + \dot{q}_n \delta p_n + p_n \delta \dot{q}_n$$

$$= -\frac{\partial L}{\partial q_n}\delta q_n + \dot{q}_n \delta p_n .$$

Hence the variation in $\mathcal{H}$ is independent of the variation in $\dot{q}$.

Now consider a special class of variations which preserve the validity of the weak equations

$$p_n \approx \frac{\partial L}{\partial \dot{q}_n} .$$

This variation then preserves the equation

$$\mathcal{H} \approx 0$$

giving

$$\delta \mathcal{H} = 0$$

Further the ϕ-equations are also preserved so that

$$\delta \phi_m = \frac{\partial \phi_m}{\partial q_n}\delta q_n + \frac{\partial \phi_m}{\partial p_n}\delta p_n = 0$$

and these are the only* conditions imposed on the variations $\delta q_n , \delta p_n$. (Any other restriction on the variations $\delta q_n , \delta p_n$ will involve also the velocities). Hence for a general variation we have,

$$\delta \mathcal{H} = \sum_{m=1}^{M} v_m \delta \phi_m$$

or

$$-\frac{\partial L}{\partial q_n}\delta q_n + \dot{q}_n \delta p_n = v_m \left(\frac{\partial \phi_m}{\partial q_n}\delta q_n + \frac{\partial \phi_m}{\partial p_n}\delta p_n \right) .$$

Equating coefficients,

$$\dot{q}_n \approx v_m \frac{\partial \phi_m}{\partial q_n} ,$$

$$\dot{p}_n \approx \frac{\partial L}{\partial q_n} \approx - v_m \frac{\partial \phi_m}{\partial p_n} ,$$

---

* Since the ϕ_m, $m = 1, \cdots, M$ form a complete set.

These are the Hamiltonian equations of motion and can be conveniently written in the Poisson bracket notation*

$$\dot{q}_n \approx v_m [q_n, \phi_m],$$
$$\dot{p}_n \approx v_m [p_n, \phi_m].$$

For any dynamical variable $\xi = \xi(p, q)$,

$$\dot{\xi} \approx v_m [\xi, \phi_m] \approx [\xi, v_m \phi_m]$$

or $$\dot{\xi} = [\xi, \mathcal{H}]$$

since $$\mathcal{H} = v_m \phi_m$$

The coefficients v_m involve the velocities.

The equations of motion of a general dynamical system with an arbitrary Lagrangian are not always consistent. For example, consider a system with one degree of freedom, denoted by the coordinate q. If we take the Lagrangian $L = q$, the equation of motion

$$\frac{d}{dt}\left(\frac{\partial L}{\partial \dot{q}}\right) = \frac{\partial L}{\partial q}$$

leads to the inconsistent result $0 \neq 1$. Hence for consistency for the dynamical scheme, certain restrictions are to be imposed. In the present case, it is necessary to have the relation

$$\phi_m \approx 0 \qquad \text{for all } t$$

so that $$\dot{\phi}_m \approx 0, \qquad m = 1, \ldots, M.$$

---

* The Poisson bracket (P.b.) of any two dynamical variables ξ, η is defined by $[\xi, \eta] = \sum_0^N \left(\frac{\partial \xi}{\partial q_n}\frac{\partial \eta}{\partial p_n} - \frac{\partial \xi}{\partial p_n}\cdot\frac{\partial \eta}{\partial q_n}\right)$.

With the Hamiltonian $\mathcal{H}$ obtained earlier, this leads to

$$\sum v_{m'} [\phi_m , \phi_{m'}] \approx 0$$

We assume* that all the Poisson brackets between any two ϕ's vanish:

$$[\phi_m , \phi_{m'}] \approx 0 , \quad m, m' = 1, \ldots , M.$$

These equations need hold only as weak equations. With this assumption, the v_m are completely arbitrary. There are then M arbitrary functions in the general solution of the equations of motion.

In addition to the ϕ-equations, the equations of motion

$$\dot{p}_n \approx \frac{\partial L}{\partial q_n}$$

may lead to further relations between q's and p's, say

$$\chi_k (q, p) \approx 0 \qquad k = 1,2,\ldots\ldots,K.$$

In the usual dynamical scheme, the equations of motion serve to fix the accelerations; if these accelerations are uniquely fixed (in terms of $q_n, \dot{q}_n$) by the equations of motion, there will be no χ-equations. In case χ-equations are present, we need the further consistency conditions

$$\sum_m v_m [\chi_k , \phi_m] \approx 0 , \qquad k = 1,\ldots\ldots,K.$$

We assume** that

$$[\chi_k , \phi_m] \approx 0$$

---

* The more general case in which some of the P. b.'s do not vanish consistency demands restrictions on the v's. For details see P.A.M. Dirac: Can. J. of Math. II, p.129 (1950).

** See the previous footnote.

and further

$$[\chi_k, \chi_{k'}] \approx 0 .$$

Consider a new Hamiltonian

$$\mathcal{H} = \sum_m v_m \phi_m + \sum_k v_k^* \chi_k \approx 0$$

with new arbitrary coefficients v_k^*. The general solution of the equations of motion now has M + K arbitrary functions. The new equations of motion are

$$\dot{\xi} = v_m [\xi, \phi_m] + v_k^* [\xi, \chi_k] .$$

All the motions allowed with the earlier Hamiltonian are allowed with the new one also. As a matter of fact, those earlier solutions are a subclass of the class of new solutions with all the v_k^* put equal to zero.

Now the ϕ 's and χ 's are treated on equal footing. Call them all $\mathcal{H}_m$'s . We may then write the new Hamiltonian as

$$\mathcal{H} = \sum^{M+K} v_m \mathcal{H}_m \approx 0$$

and all the $\mathcal{H}_m$'s have zero P.b 's with each other.

We may now pass to the quantum theory by replacing the weak equations $\mathcal{H}_m \approx 0$ by Schrödinger equations:

$$\mathcal{H}_m \Psi = 0 .$$

When there exist χ equations, the transition to the new Hamiltonian is necessary before one can pass to the quantum theory.

Lecture 18.

To illustrate the theory developed in the last lectures we shall consider the example of a scalar field $V(x_0, x_1, x_2, x_3)$. The action density is given by

$$L = \frac{1}{2} \frac{\partial V}{\partial x_\mu} \frac{\partial V}{\partial x^\mu} - \frac{1}{2} m^2 V^2$$
$$= \frac{1}{2} V_\mu V^\mu - \frac{1}{2} m^2 V^2 \tag{1}$$

where $V^\mu \equiv \frac{\partial V}{\partial x_\mu}$ and m is a constant. By variation of the action integral

$$I = \int L \, d^4 x \,, \quad d^4 x = dx_0 \, dx_1 \, dx_2 \, dx_3 \tag{2}$$

we obtain

$$\delta I = \int (V_\mu \delta V^\mu - m^2 V \delta V) \, d^4 x$$
$$= - \int \left(\frac{\partial V_\mu}{\partial x_\mu} + m^2 V \right) \delta V \, d^4 x, \tag{3}$$

which is zero if

$$\frac{\partial^2 V}{\partial x_\mu \partial x^\mu} + m^2 V = 0. \tag{4}$$

To designate the points of a curved three dimensional space-like surface we introduce a system of curvilinear coordinates in it. Let u_r, $r=1, 2, 3$ be the coordinates of a point in the surface. Let us denote by y_μ the four space-time coordinates of a point in the surface. The surface will then be fixed by

$$y_\mu = y_\mu(u) \tag{5}$$

being given, for all u.

We define the metric γ^{rs} in the surface by writing for the displacement ds in the surface

$$ds^2 = \gamma^{rs}\, du_r\, du_s . \tag{6}$$

Since also

$$\begin{aligned}(ds)^2 &= g^{\mu\nu}\, dy_\mu(u)\, dy_\nu(u) \\ &= g^{\mu\nu}\, y_\mu^r\, y_\nu^s\, du_r\, du_s ,\end{aligned} \tag{7}$$

we have

$$\gamma^{rs} = y_\mu^r\, y_\nu^s\, g^{\mu\nu} = y_\mu^r\, y^{\mu s} . \tag{8}$$

Here $y_\mu^r \equiv \frac{\partial y_\mu}{\partial u_r} \cdot \gamma^{rs}$ satisfies the relation (with γ_{rs} = cofactor of $\gamma^{rs}/\det \gamma^{rs}$)

$$\gamma_{rp}\, \gamma^{rs} = \delta_p{}^s \tag{9}$$

Since the surface is space-like, $(ds)^2$ is negative and so the determinant of γ^{rs} must be negative. We put

$$\det |\gamma^{rs}| = -\Gamma^2 , \tag{10}$$

Γ being a real number. An element of the surface is now given by

$$\Gamma\, du_1\, du_2\, du_3 \equiv \Gamma\, d^3u$$

We shall also make use of the unit normal to the surface $l_\mu = l_\mu(u)$ which satisfies

$$\left.\begin{aligned} l_\mu\, y^{\mu r} &= 0 , \\ l_\mu\, l^\mu &= 1 , \quad l_0 > 1 \end{aligned}\right\} \tag{11}$$

The field variables on the surface are now fixed by $V = V(u)$.

The q 's of our theory will be $y_\mu(u)$ and $V(u)$ for all u -values. They will be regarded as functions of an independent variable τ, so that the velocities are $\dot{y}_\mu, \dot{V}$, the dot denoting differentiation with respect to τ.

The action between two neighbouring surfaces $y_\mu(u)_\tau$ and $y_\mu(u)_{\tau+\delta\tau}$ is

$$\mathcal{L}\,\delta\tau = \int L\, l_\mu \dot{y}^\mu\, \delta\tau\, \Gamma d^3u, \tag{12}$$

whence

$$\mathcal{L} = \int L\, \dot{y}_e\, \Gamma d^3u = \tfrac{1}{2}\int (V_\mu V^\mu - m^2 V^2)\dot{y}_e\, \Gamma d^3u, \tag{13}$$

where we have written $\dot{y}_e = l_\mu \dot{y}^\mu$. This is the normal derivative of y_μ.

In general any vector A_μ has a normal component $A_e = A_\mu l^\mu$ and a tangential component $A^r = A_\mu y^{\mu r}$. Then it is easily seen that

$$A_\mu = A_e l_\mu + A^r y_{\mu r}. \tag{14}$$

Now

$$\dot{V} = V_\mu \dot{y}^\mu = V_e \dot{y}_e + V^r \dot{y}_r, \tag{15}$$

where

$$\dot{y}_r = \dot{y}^\mu y_{\mu r} = \dot{y}^\mu \frac{\partial y_\mu}{\partial u_s}\gamma_{rs}. \tag{16}$$

From (14) and (15) we get

$$V_\mu = \frac{\dot{V} - V^r \dot{y}_r}{\dot{y}_e} l_\mu + V^r y_{\mu r}. \tag{17}$$

V_μ is thus expressed in terms of coordinates and velocities only. Substituting this in the expression (13) for $\mathcal{L}$ we get the Lagrangian in the standard form.

To define the conjugate momentum we cannot now use the definition

$$p_n = \frac{\partial \mathcal{L}}{\partial \dot{q}_n}, \tag{18}$$

for now there is a continuous infinity of velocities and $\mathcal{L}$ is a functional of them. Now (18) can be written as

$$\delta \mathcal{L} = p_n \, \delta \dot{q}_n , \tag{19}$$

for a general variation of the velocities.

This relation suggests the following definition for the momentum conjugate to q in the continuous case:

$$\delta \mathcal{L} = \int p \, \delta \dot{q} \; d^3 u , \tag{20}$$

for a general variation of the velocities.

Varying velocities only we obtain from (13)

$$\delta \mathcal{L} = \int \{ V^{\mu} \delta V_{\mu} \dot{y}_{e} + \tfrac{1}{2} (V_{\mu} V^{\mu} - m^2 V^2) \, \delta \dot{y}_{e} \} \Gamma \, d^3 u . \tag{21}$$

From (17)

$$\begin{aligned} \delta V_{\mu} &= l_{\mu} \left\{ \frac{\delta \dot{V} - V^{\nu} \delta \dot{y}_{\nu}}{\dot{y}_{e}} - \frac{\dot{V} - V^{\nu} \dot{y}_{\nu}}{\dot{y}_{e}^{2}} \, \delta \dot{y}_{e} \right\} \\ &= l_{\mu} \frac{\delta \dot{V} - V_{e} \, \delta \dot{y}_{e} - V^{\nu} \delta \dot{y}_{\nu}}{\dot{y}_{e}} = l_{\mu} \frac{\delta \dot{V} - V_{\rho} \, \delta \dot{y}^{\rho}}{\dot{y}_{e}} . \end{aligned} \tag{22}$$

Substituting in (21) we get

$$\delta \mathcal{L} = \int \{ V_{e} (\delta \dot{V} - V_{\mu} \delta \dot{y}^{\mu}) + \tfrac{1}{2} (V_{\mu} V^{\mu} - m^2 V^2) \, l_{\sigma} \, \delta \dot{y}^{\sigma} \} \Gamma \, d^3 u . \tag{23}$$

Now $\delta \mathcal{L}$ must be expressible as

$$\delta \mathcal{L} = \int (u \, \delta \dot{V} + w^{\sigma} \, \delta \dot{y}_{\sigma}) \, d^3 u , \tag{24}$$

where u is the momentum conjugate to V and w^{σ} that to y_{σ}. Comparing (23) and (24) we obtain

$$u \approx V_e \Gamma , \tag{25a}$$

$$w^\sigma \approx -V_e V^\sigma \Gamma + \frac{1}{2}(V_\mu V^\mu - m^2 V^2) e^\sigma \Gamma . \tag{25b}$$

These satisfy the P.B. relations

$$\left. \begin{aligned} [V, u'] &= \delta_3(u-u') , \\ [y_\mu , w'_\nu] &= g_{\mu\nu} \delta_3(u-u') . \end{aligned} \right\} \tag{26}$$

The tangential and the normal parts of (25b) are, respectively,

$$w^r \approx -V_e \Gamma V^r \approx -U V^r \tag{27}$$

$$\begin{aligned} w_e &\approx -V_e^2 \Gamma - \frac{1}{2}(V_e^2 + V_r V^r - m^2 V^2)\Gamma \\ &\approx -\frac{1}{2}\frac{U^2}{\Gamma} - \frac{1}{2} V_r V^r \Gamma - \frac{1}{2} m^2 V^2 \Gamma . \end{aligned} \tag{28}$$

Hence we obtain the Hamiltonians

$$\mathcal{H}^r = w^r + U V^r \approx 0 \tag{29}$$

$$\mathcal{H}_e = w_e + \frac{1}{2}\left(\frac{U^2}{\Gamma} - V_r V^r \Gamma + m^2 V^2 \Gamma\right) . \tag{30}$$

Since there are no χ equations in the present case these are the only Hamiltonians. The P.B. of these vanish. The general Hamiltonian will be

$$\mathcal{H}_{gen} = \int (v_r \mathcal{H}^r + v_e \mathcal{H}_e) d^3u \approx 0 , \tag{31}$$

the v's being arbitrary coefficients.

We may write (31) as

$$\mathcal{H}_{gen} = \int v_\mu \mathcal{H}^\mu d^3u \approx 0 . \tag{32}$$

Then $\dot{y}_\nu = [y_n, \mathcal{H}_{gen}] = [y_\nu, \int v'_\mu \mathcal{H}'^\mu d^3u']$

$$\approx \int v'_\mu [y_\nu, \mathcal{H}'^\mu] d^3u'. \tag{33}$$

Now $\mathcal{H}_\mu = w_\mu +$ something independent of w_μ, so

$$[y_\nu, \mathcal{H}'^\mu] = \delta_\nu^\mu \, \delta_3(u - u'), \tag{34}$$

and (33) reduces to

$$\dot{y}_\nu = v_\nu. \tag{35}$$

Hence we may write

$$\mathcal{H}_{gen} = \int \dot{y}_\mu \mathcal{H}^\mu d^3u = \int \dot{y}_r \mathcal{H}^r d^3u + \int \dot{y}_e \mathcal{H}_e d^3u. \tag{36}$$

$\mathcal{H}_r$ gives a change in the parametrization of the surface, since it produces motions in the surface. $\mathcal{H}_e$ leads to motions of points of the surface normal to it.

Curved surfaces are inconvenient to use when practical calculations are to be carried out. They are necessary only to have a formal relativistic theory. For practical calculations we shall revert to flat surfaces t = const., referring to a single observer.

Then

$$\dot{y}_r = 0, \ \dot{y}_e = 1, \quad \dot{y}_0 = \text{constant (independent of the } u\text{'s)} \qquad \Gamma = 1, \tag{37}$$

and

$$\mathcal{H}_{gen} = \dot{y}_0 \int \mathcal{H}_0 \, d^3u$$

$$= \dot{y}_0 \int \{ w_0 + \tfrac{1}{2}(u^2 - V_r V^r + m^2 V^2) \} d^3u$$

$$= \dot{y}_0 \, (W + H) \, \varkappa_0 \qquad (38)$$

with $\quad W = \int w_0 \, d^3u \; , \; H = \tfrac{1}{2}\int (u^2 - V_r V^r + m^2 V^2) d^3u \; . \qquad (39)$

W is the momentum conjugate to the single variable y_0 required to fix the time of the flat surface. H is the usual expression for the energy.

Lecture 19.

We now discuss the quantisation of the generalised Hamiltonian theory of the scalar field V. Since we are working out the properties of a definite dynamical system we restrict ourselves to flat surfaces with the advantage of considerable simplification. The dynamical variables $V_{(u)}$, $U_{(u)}$ form an infinite set, one for each point u. In the classical theory these are Poisson bracket relations connecting them which are replaced by quantum Poisson brackets (commutation brackets) in the quantised theory.

Consider the question of representations. Since the V's for different u-values all commute we may take the wave function ψ to be

$$\psi = \psi(V)$$

Such a function of an infinite number of variables is called a functional; the representation is hence by functionals. Such a representation, however, leads to considerable mathematical difficulties and is not useful physically, because we do not usually observe the V's.

So we pursue another line of attack and resolve the field at any instant into its (3-dimensional) Fourier components, each representing a harmonic oscillator. For this purpose, we introduce a 3-dimensional vector $\underline{k}$ (with components k_1, k_2, k_3) to specify the Fourier component in 3 dimensions. The Fourier components $\xi_{\underline{k}}$ will in general be complex. With the notation

$$\underline{k}\cdot\underline{u} = k_1 u_1 + k_2 u_2 + k_3 u_3$$

the Fourier decomposition is given by

$$V(\underline{u}) = \int\left\{ \xi_{\underline{k}}\, \bar{e}^{i\underline{k}\underline{u}} + \bar{\xi}_{\underline{k}}\, e^{i\underline{k}\cdot\underline{u}} \right\} d^3\underline{k}$$

$$= \int (\xi_{\underline{k}} + \bar{\xi}_{-\underline{k}})\, \bar{e}^{i\underline{k}\cdot\underline{u}}\, d^3\underline{k}$$

From the last equality, it is easily seen that the decomposition is not unique and fixes only $(\xi_{\underline{k}} + \bar{\xi}_{-\underline{k}})$. To make the decomposition unique we need another equation, which will be given below.

We proceed as follows: differentiate V obtaining

$$\frac{\partial V}{\partial u_r} = -i\int k_r\,(\xi_{\underline{k}} + \bar{\xi}_{-\underline{k}})\, \bar{e}^{i\underline{k}\cdot\underline{u}}\, d^3\underline{k}$$

so that

$$\int \frac{\partial V}{\partial u_r}\frac{\partial V}{\partial u_r}\, d^3u = -\iiint k_r k'_s\,(\xi_{\underline{k}} + \bar{\xi}_{-\underline{k}})(\xi_{\underline{k}'} + \bar{\xi}_{-\underline{k}'}) \times \bar{e}^{i\underline{k}\cdot\underline{u} - i\underline{k}'\cdot\underline{u}}\, d\underline{k}\, d\underline{k}'\, d^3u$$

Now

$$\int \bar{e}^{i\underline{a}\cdot\underline{u}}\, d^3u = 8\pi^3\,\delta_3(\underline{a})$$

so that

$$\int \frac{\partial V}{\partial u_r}\frac{\partial V}{\partial u_r}\, d^3u = 8\pi^3\int k_r^2\,(\xi_{\underline{k}} + \bar{\xi}_{-\underline{k}})(\xi_{-\underline{k}} + \bar{\xi}_{\underline{k}})\, d^3k$$

Similarly

$$\int m^2 V^2\, d^3u = 8\pi^3\int m^2(\xi_{\underline{k}} + \bar{\xi}_{-\underline{k}})(\xi_{-\underline{k}} + \bar{\xi}_{\underline{k}})\, d^3k\,.$$

Assume

$$U(u) = i\int (k^2+m^2)^{\frac{1}{2}}\,(\xi_{\underline{k}} - \bar{\xi}_{-\underline{k}})\, \bar{e}^{i\underline{k}\cdot\underline{u}}\, d^3\underline{k}$$

This equation fixes $(\xi_{\underline{k}} - \bar{\xi}_{-\underline{k}})$ uniquely and coupled with the above decomposition of $V(u)$ fixes $\xi_{\underline{k}}$ uniquely.

With U given as above, we have,

$$\int U^2 d^3u = -8\pi^3 \int (k^2+m^2)(\xi_{\underline{k}} - \bar{\xi}_{-\underline{k}})(\xi_{-\underline{k}} - \bar{\xi}_{\underline{k}}) d^3k .$$

The Hamiltonian hence becomes

$$H = \frac{1}{2}\int \left(V^2 + \frac{\partial U}{\partial u_r}\frac{\partial U}{\partial u_r} + m^2 \right) d^3u = 16\pi^3 \int (k^2+m^2)\, \xi_{\underline{k}}\, \bar{\xi}_{-\underline{k}}\, d^3k$$

Inverting the Fourier decompositions, one obtains

$$\xi_{\underline{k}} + \bar{\xi}_{-\underline{k}} = \frac{1}{8\pi^3} \int U e^{i\underline{k}\cdot\underline{u}} d^3u$$

$$(k^2+m^2)^{\frac{1}{2}} (\xi_{\underline{k}} - \bar{\xi}_{-\underline{k}}) = \frac{-i}{8\pi^3} \int V e^{i\underline{k}\cdot\underline{u}} d^3u$$

Further, since U and V are conjugate to each other, they satisfy the Poisson bracket relations

$$[V, V'] = 0 \quad , \quad [U, U'] = 0$$

$$[V, U'] = \delta_3(\underline{u} - \underline{u}')$$

where, as before, V' refers to $V(u')$ etc. Using these relations together with the Fourier inverses, we obtain,

$$[\xi_{\underline{k}} + \bar{\xi}_{-\underline{k}} , \xi_{\underline{k}'} + \bar{\xi}_{-\underline{k}'}] = 0$$

$$[\xi_{\underline{k}} - \bar{\xi}_{-\underline{k}} , \xi_{\underline{k}'} - \bar{\xi}_{-\underline{k}'}] = 0$$

$$[\xi_{\underline{k}} + \bar{\xi}_{-\underline{k}} , \xi_{\underline{k}'} - \bar{\xi}_{-\underline{k}'}]$$

$$= \frac{-i}{64\pi^6} \frac{1}{(k'^2+m^2)^{\frac{1}{2}}} \iint [U, U'] e^{i\underline{k}\cdot\underline{u} - i\underline{k}'\cdot\underline{u}'} d^3u\, d^3u'$$

$$= \frac{-i\,(k^2+m^2)^{1/2}}{8\pi^3} \delta_3(\underline{k}+\underline{k}')$$

These relations are equivalent to the commutation relations

$$[\xi_{\underline{k}}, \xi_{\underline{k}'}] = 0$$

and

$$[\xi_{\underline{k}}, \bar{\xi}_{-\underline{k}'}] = \frac{i}{16\pi^3} \frac{1}{(k^2+m^2)^{1/2}} \delta_3(\underline{k}+\underline{k}')$$

Changing $-\underline{k}'$ to $\underline{k}'$, the last commutator can be written

$$[\xi_{\underline{k}}, \bar{\xi}_{\underline{k}'}] = \frac{i}{16\pi^3} \frac{1}{(k^2+m^2)^{1/2}} \delta_3(\underline{k}-\underline{k}')$$

The commutation relations obeyed by the ξ are the same, (upto a factor) as the ones obeyed by the variables representing a set of harmonic oscillators in Fock's representation. In this representation, we had, for each oscillator a variable η_r and its complex conjugate $\bar{\eta}_r$ which satisfy

$$\eta_r \eta_s - \eta_s \eta_r = 0$$

$$\bar{\eta}_r \eta_s - \eta_s \bar{\eta}_r = \delta_{rs}$$

The standard ket $|S\rangle$ is chosen to represent the normal state of each oscillat

$$\bar{\eta}_r |S\rangle = 0$$

Any ket $|P\rangle$ can be expressed as a power series in the η_r multiplying the standard ket:

$$|P\rangle = \sum c_{n_1 n_2 \ldots} \eta_1^{n_1} \eta_2^{n_2} \cdots |S\rangle .$$

The set of oscillators is equivalent to an assembly of bosons, with each oscillator corresponding to one state for the bosons. To complete the identification let the states of the bosons be $\alpha_a, \alpha_b, \alpha_c, \ldots$ and equate the ket

$$\eta_1^{n_1} \eta_2^{n_2} \ldots |S\rangle$$

to the symmetrised ket

$$\sum_p |\alpha_a \alpha_b \alpha_c \ldots\rangle$$

apart from a normalisation factor, where $n_1, n_2, n_3, \ldots$ are the numbers of bosons in the states $\alpha_1, \alpha_2, \alpha_3 \ldots$ respectively. We may write including the normalisation factor

$$\eta_1^{n_1} \eta_1^{n_2} \ldots |S\rangle = \frac{1}{\{(\sum n_a)!\}^{1/2}} \sum_p |\alpha_a \alpha_b \ldots\rangle$$

since the squared length of each side is now $n_1! \, n_2! \ldots$. This shows that the oscillators and the bosons are simply two ways of looking at the same thing.

In the present case we have an infinite number of oscillators, one for each $\underline{k}$ -value. If we denote the Fock variables by $\eta_{\underline{k}}$, we should have

$$\eta_{\underline{k}} \eta_{\underline{k}'} - \eta_{\underline{k}'} \eta_{\underline{k}} = 0$$
$$\bar{\eta}_{\underline{k}} \eta_{\underline{k}'} - \eta_{\underline{k}'} \bar{\eta}_{\underline{k}} = \delta_3(\underline{k} - \underline{k}')$$

by a natural generalisation to the case of a continuous infinity of oscillators. Notice that the Kronecker delta δ_{rs} of the discrete case is replaced by the delta function $\delta_3(\underline{k} - \underline{k}')$. We must now put

$$\xi_{\underline{k}} = \frac{\hbar^2}{4\pi^{3/2}} \frac{1}{(k^2 + m^2)^{1/4}} \eta_{\underline{k}}$$

In the present case, an expansion of the general ket into a power series of the η 's multiplying the standard ket, is replaced by an expansion of the form

$$|P\rangle = \{c_0 + \int c_{\underline{k}}\, \eta_{\underline{k}}\, d^3\underline{k} + \iint c_{\underline{k}\,\underline{k}'}\, \eta_{\underline{k}}\, \eta_{\underline{k}'}\, d^3\underline{k}\, d^3\underline{k}' + \cdots \}\, |S\rangle$$

with each term of the power series replaced by a (multiple) integral in the $\underline{k}$ -space. The coefficient of $|S\rangle$ is now a functional of $\eta_{\underline{k}}$. To interpret this functional we note that the first term corresponds to no bosons being present. For the second term, we note that $\eta_{\underline{k}}\, |S\rangle$ is a ket which is not normalisable (infinite in length); however

$$|c_{\underline{k}}|^2\, dk_1\, dk_2\, dk_3$$

gives the probability of one boson being in a state with a $\underline{k}$ lying in the volume element $dk_1\, dk_2\, dk_3$ in the $\underline{k}$ space. The third term corresponds similarly to two bosons being present, and so on.

The order of the factors $\xi_{\underline{k}}\, \bar{\xi}_{\underline{k}}$ in the Hamiltonian H should be so chosen in the quantum theory as to yield vanishing zero point energy. This requires that the $\bar{\xi}$'s shall all be to the right of the ξ 's. Since

$$\xi_{\underline{k}}\, \bar{\xi}_{\underline{k}}\, |S\rangle = 0 \quad \text{, we now get} \quad H\, |S\rangle = 0$$

With U and V not commuting, we have

$$\frac{1}{2}\int \left(U^2 + \frac{\partial V}{\partial u_r}\frac{\partial V}{\partial u_r} + m^2\right) d^3u$$

$$= 16\pi^3 \int (k^2+m^2)^{\frac{1}{2}} . \, \frac{1}{2}\left(\xi_{\underline{k}}\, \bar{\xi}_{\underline{k}} + \bar{\xi}_{\underline{k}}\, \xi_{\underline{k}}\right) d^3k$$

as can easily be seen from symmetry considerations . But

$$\frac{1}{2}\left(\xi_{\underline{k}}\, \bar{\xi}_{\underline{k}} + \bar{\xi}_{\underline{k}}\, \xi_{\underline{k}}\right) = \frac{1}{2}\left[2\, \xi_{\underline{k}}\, \bar{\xi}_{\underline{k}} + \left(\bar{\xi}_{\underline{k}}\, \xi_{\underline{k}} - \xi_{\underline{k}}\, \bar{\xi}_{\underline{k}}\right)\right]$$

$$= \xi_{\underline{k}}\, \bar{\xi}_{\underline{k}} + \infty .$$

Hence this leads to an infinite zero point energy. This Hamiltonian $\left[i.e., \frac{1}{2}\int \left[\dot{V}^2 + \frac{\partial V}{\partial u_r}\frac{\partial V}{\partial u_r} + m^2\right] d^3u \right]$ is thus not suitable for the quantum theory. It is not convenient to try to subtract the terms giving rise to the zero point energy, without working with the Fourier components. Since we need the Fourier components anyway to get a useful physical interpretation, complication need not concern us any further.

Lecture 20

We saw in the last lecture that an assembly of bosons could be represented as a system of harmonic oscillators and we obtained a suitable representation for dealing with it. We shall now introduce a corresponding representation for an assembly of fermions.

We have now to insure that there cannot exist more than one fermion in the same state. Let η_a , $\bar{\eta}_a$ be the emission and absorption operators respectively, for a fermion in state a , considering discrete states first. We must then have

$$\eta_a^2 = 0 \qquad (1a)$$

$$\bar{\eta}_a^2 = 0 \qquad (1b)$$

which express the impossibility of having two particles in the same state.

Let $|S\rangle$ denote the standard ket representing a state with no particles at all. Then

$$\bar{\eta}_a |S\rangle = 0 \qquad (2)$$

and $\eta_a |S\rangle$ is the ket representing a state with but one particle in state a . With suitable normalization of η_a , $\bar{\eta}_a$ we then have

$$\bar{\eta}_a \eta_a |S\rangle = |S\rangle \qquad (3)$$

From (3) and (2) we get

$$\eta_a \bar{\eta}_a \eta_a |S\rangle = \eta_a |S\rangle , \qquad (4a)$$

$$\eta_a \bar{\eta}_a |S\rangle = 0 \qquad (4b)$$

Thus $\eta_a \bar{\eta}_a$ applied to a ket with just one particle in state a is equivalent to 1, and to a ket with no particle in state a to 0. This statement is still valid when there are particles in other states. We may thus put

$$\eta_a \bar{\eta}_a = n_a , \qquad (5)$$

n_a being the number of particles in the state a.

Similarly from

$$\bar{\eta}_a \eta_a |S\rangle = |S\rangle , \qquad (6a)$$

$$\bar{\eta}_a \eta_a \eta_a |S\rangle = 0 , \qquad (6b)$$

we see that

$$\bar{\eta}_a \eta_a = 1 - n_a . \qquad (7)$$

From (5) and (7) we have.

$$\eta_a \bar{\eta}_a + \bar{\eta}_a \eta_a = 1 \qquad (8)$$

We shall now study the commutation relations for operators $\eta , \bar{\eta}$ for different states a, b. A general ket with fermions in states $a, b, c, \ldots$ (all different) can be represented as

$$\eta_a \eta_b \eta_c \cdots |S\rangle . \qquad (9)$$

The same can also be represented by the wave function $\psi(\alpha_a, \alpha_b, \alpha_c, \ldots)$ where α 's are dynamical variables which have specified values for a given state of the fermion. This wave function can be equated to (9). Since ψ is antisymmetric in the α's, we require the η 's to anticommute:

$$\eta_a \eta_b + \eta_b \eta_a = 0 . \qquad (10)$$

Since η_a is an observable, independent of $\eta_b, \bar{\eta}_b$ for b different from a, we should expect to have

$$\eta_b n_a = n_a \eta_b \quad , \quad a \neq b \qquad (11)$$

Henée

$$\eta_b \bar{\eta}_a = - \bar{\eta}_a \eta_b \quad , \quad a \neq b \, . \qquad (12)$$

Thus the commutation relations of the operators $\eta_a, \bar{\eta}_a$ are

$$\eta_a \eta_b + \eta_b \eta_a = 0 \, , \qquad (13)$$

$$\eta_a \bar{\eta}_b + \bar{\eta}_b \eta_a = \delta_{ab} \, ; \qquad (13a)$$

the latter can be split into

$$\eta_a \bar{\eta}_a = n_a \, , \qquad (13b)$$

$$\bar{\eta}_a \eta_a = 1 - n_a \, . \qquad (13c)$$

These are closely similar to the commutation relations for boson operators

$$\eta_a \eta_b - \eta_b \eta_a = 0 \, ,$$

$$\bar{\eta}_a \eta_b - \eta_b \bar{\eta}_a = \delta_{ab} \, ,$$

$$\eta_a \bar{\eta}_a = n_a \, ,$$

$$\bar{\eta}_b \eta_a = 1 + n_a \, .$$

In both cases a general ket $|P\rangle$ can be expressed as

$$|P\rangle = (\text{power series in } \eta\text{'s}) \, |S\rangle \, . \qquad (14)$$

In the case of fermions the power series must not involve powers higher than one of any η_a.

This formalism can be immediately generalized to the case of a continuous range of a-values by replacing the Kronecker symbol δ_{ab} by the

Dirac δ function $\delta(a-b)$

We shall now treat the electromagnetic field without interaction with particles according to the general theory developed in the previous lectures. This example presents certain new features as a consequence of the possibility of gauge transformations. We shall for simplicity's sake consider flat surfaces, though the treatment could be taken over to the case of general curved surfaces by straightforward methods.

One must have an action density $\mathcal{L}$ and an action principle

$$\delta \int \mathcal{L}\, d^4x = 0 \tag{15}$$

Suppose $\mathcal{L}$ is a function of a field quantity K and its first derivatives $K^\mu = \frac{\partial K}{\partial x_\mu}$ Then (15) is

$$0 = \int \left\{ \frac{\partial \mathcal{L}}{\partial K} \delta K + \frac{\partial \mathcal{L}}{\partial K^\mu} \delta K^\mu \right\} d^4x$$

$$= \int \left\{ \frac{\partial \mathcal{L}}{\partial K} - \frac{\partial}{\partial x_\mu} \frac{\partial \mathcal{L}}{\partial K^\mu} \right\} \delta K\, d^4x$$

whence

$$\frac{\partial \mathcal{L}}{\partial K} - \left(\frac{\partial \mathcal{L}}{\partial K^\mu}\right)^\mu = 0, \tag{16}$$

These are the field equations.

For our present example we choose

$$\mathcal{L} = -\tfrac{1}{4} F_{\mu\nu} F^{\mu\nu}$$

where $F^{\mu\nu}$ is the six vector giving the field; in terms of the potentials A^μ

$$F^{\mu\nu} = \frac{\partial A^{\nu}}{\partial x_{\mu}} - \frac{\partial A^{\mu}}{\partial x_{\nu}}$$
$$= A^{\nu\mu} - A^{\mu\nu}. \qquad (18)$$

Taking K to be A^{σ} we see that the field equations (16) are now

$$\tfrac{1}{2}\left(F_{\mu\nu} \frac{\partial F^{\mu\nu}}{\partial A^{\sigma\rho}}\right)^{\rho} = 0,$$

or

$$\tfrac{1}{2}\left\{F_{\mu\nu}\left(\delta_{\sigma}^{\nu}\delta_{\rho}^{\mu} - \delta_{\sigma}^{\mu}\delta_{\rho}^{\nu}\right)\right\}^{\rho} = 0,$$

or

$$F_{\rho\sigma}{}^{\rho} = 0 \qquad (19)$$

These are just the Maxwell field equations, so our choice for the action density is correct.

Now we must get the Lagrangian and pass to the Hamiltonian formalism by the standard method. We define our flat surfaces by y_0 = constant, and introduce an independent variable τ, different from y_0. The Lagrangian L is defined by

$L\,\delta\tau$ = amount of action between the surfaces $y_0(\tau)$ and $y_0(\tau + \delta\tau)$

Introducing the cartesian coordinates u_1, u_2, u_3 in the surface we have

$$L = \dot{y}_0 \int \mathcal{L}\, d^3u$$
$$= -\tfrac{1}{4}\,\dot{y}_0 \int \left(F_{rs}F^{rs} + 2F_{r0}F^{r0}\right) d^3u, \qquad (20)$$

where suffices r, s run through 1,2,3. Now

$$F^{r0} = \frac{\partial A^0}{\partial x_r} - \frac{\partial A^r}{\partial x_0}$$
$$= A^{0r} - \frac{\dot{A}^r}{\dot{y}_0} , \qquad (21)$$

where the dot denotes differentiation with respect to τ. Substituting (21) into (20) we get

$$L = -\dot{y}_0 \int \{ \tfrac{1}{4} F_{rs} F^{rs} + \tfrac{1}{2} (A^{0r} - \frac{\dot{A}^r}{\dot{y}_0})(A_{0r} - \frac{\dot{A}_r}{\dot{y}_0}) \} d^3u \qquad (22)$$

which is a function of coordinates and velocities and is homogeneous and of the first degree in the velocities.

To introduce the momenta we vary the velocities, and pick out the coefficients of the velocities in the expression for δL. We have

$$\delta L = -\delta \dot{y}_0 \int \tfrac{1}{4} F_{\mu\nu} F^{\mu\nu} + \dot{y}_0 \int (A_{0r} - \frac{\dot{A}_r}{\dot{y}_0})(\frac{\delta \dot{A}^r}{\dot{y}_0} - \frac{\dot{A}^r}{\dot{y}_0^2} \delta \dot{y}_0) \times d^3u \qquad (23)$$

Let W be the conjugate momentum to y_0 and B_μ to A^μ; then

$$\delta L = W \delta \dot{y}_0 + \int B_\mu \, \delta \dot{A}^\mu \, d^3u . \qquad (24)$$

Comparing (23) and (24) we get

$$B_r \approx F_{r,0} \qquad (25a)$$
$$B_0 \approx 0 \qquad (25b)$$
$$W \approx -\tfrac{1}{4} \int \{ F_{\mu\nu} F^{\mu\nu} + F_{r0} \frac{\dot{A}^r}{\dot{y}_0} \} d^3u . \qquad (25c)$$

For these we have the P.b. relations

$$[y_0\, W] = 1, \qquad (26a)$$

$$[A^\mu, B'_\nu] = \delta_\nu^\mu\, \delta_3(u-u')\, ; \qquad (26b)$$

the P.bs. between other pairs are zero.

From (25) we shall now extract equations involving coordinates and momenta only. They will be our ϕ equations. Equation (25b) is itself a ϕ equation. The velocities occurring in (25c) are removed by using (25a) and

$$\frac{\dot{A}^r}{\dot{y}_0} = A^{r0} = F^{0r} + A^{0r} = -B^r + A^{0r}.$$

we get $W \approx -\int \{ \frac{1}{4} F_{rs} F^{rs} + \frac{1}{2} B_r B^r + B_r (A^{0r} - B^r) \} d^3u$ Hence

$$W + \int \{ \tfrac{1}{4} F_{rs} F^{rs} - \tfrac{1}{2} B_r B^r - A^0\, B_r{}^r \} d^3u \approx 0, \qquad (27)$$

in which we have made use of a partial integration to get the last term. One easily checks that these ϕ 's have vanishing P.b.

Our general Hamiltonian will now be a linear function of (25b) and (27) with arbitrary coefficients :

$$\mathcal{H}_{gen} = v\,[\, W + \int \{ \tfrac{1}{4} F_{rs} F^{rs} - \tfrac{1}{2} B_r B^r - A^0\, B_r{}^r \} d^3u\,]$$
$$+ \int \lambda\, B_0\, d^3u, \qquad (28)$$

where v and $\lambda = \lambda(u)$ are the arbitrary coefficients.

Using (26) we obtain

$$\dot{y}_0 = [y_0, \mathcal{H}_{gen}] \approx v_0$$

which fixes the coefficient v. To get a meaning for λ we put $v = 0$, which means that the surface does not move. The dynamical variables can, however, still change, for we have

$$\dot{\xi} = [\xi, \int \lambda' B_0' \, d^3u']$$
$$\approx \int \lambda' [\xi, B_0'] \, d^3u' . \qquad (29)$$

This equation shows that the only dynamical variable affected will be A_0, and

$$\dot{A}_0 = \lambda . \qquad (30)$$

The coefficient of B_0 in (28) thus gives the rate of change in A_0.

The Hamiltonian (28) is not the most general one in the present case, for now we also have χ equations. Putting $\sigma = 0$ in the field equations (19) we obtain

$$F_{r0}{}^{r} \approx 0$$

and so, using (25a)

$$B_r{}^r \approx 0 . \qquad (31)$$

This is a χ equation. This χ has zero P.bs. with the ϕ's, as can easily be verified. Taking this into account we obtain the new Hamiltonian

$$\mathcal{H}_{new} = \mathcal{H}_{prev} + \int \kappa \, B_r{}^r \, d^3u , \qquad (32)$$

$\kappa = \kappa(u)$ being arbitrary.

This new Hamiltonian will allow more general equations of motion then the previous one. To see the nature of this generality we put

$$\dot{y}_0 = 0 , \quad \lambda = 0 .$$

Then

$$\dot{\xi} = \int K' [\xi, B_r^{\,r'}]\, d^3u\,, \qquad (33)$$

where the dash on the superscript r denotes that it refers to the variable of differentiation u'_r. This shows that the only dynamical variable varying now will be A^s. We have

$$[A^s, B_r^{\,r'}] = [A^s, B'_r]^{r'} = \delta_r^{\,s} (\delta_3(u-u'))^{r'}$$

$$= -(\delta_3(u-u'))^{s'}. \qquad (34)$$

From equations (33) and (34) we now obtain

$$\dot{A}^s = \int K' \,\delta_3^{\,s}(u-u')\, d^3u'$$

$$= \frac{\partial K}{\partial u_s}. \qquad (35)$$

This is a gauge transformation.

A general gauge transformation is

$$A^\mu \longrightarrow A^\mu + \frac{\partial S}{\partial x_\mu}.$$

If we consider it applied at one particular instant of time, then it can be considered as made up of two parts

$$A^0 \longrightarrow A^0 + \frac{\partial S}{\partial x_0}, \qquad (A)$$

and

$$A^r \longrightarrow A^r + \frac{\partial S}{\partial x_r}, \qquad (B)$$

since $\frac{\partial S}{\partial x_0}$ and S are arbitrary at one instant of time. The gauge transformation (A) is already produced by the term involving λ in $\mathcal{H}_{pr}$

whereas (B) is taken care of by the κ term in $\mathcal{H}_{new}$. The Hamiltonian (32) is thus more physical, since it automatically allows all the gauge transformations, to which a theory of the electromagnetic field should be invariant.

Lecture 21.

The new Hamiltonian of the free electromagnetic field can be written

$$H_{new} = \dot{y}_0 \{ W + \int (\tfrac{1}{4} F_{rs} F^{rs} - \tfrac{1}{2} B_r B^r) d^3u \} + \lambda \int B_0 d^3u + \int \kappa B_r{}^r d^3u \approx 0 .$$

Here we have absorbed the term $-A_0 B_r{}^r$ in the first integral into the last integral, which we can do since κ is arbitrary.

A_0 has now disappeared from the Hamiltonian. Furtner, B_0 occurs in the Hamiltonian only in the term $\int \lambda B_0 d^3u$, whose effect on the equations of motion is merely to allow A_0 to vary arbitrarily; also we have

$$B_0 \approx 0 .$$

Hence we can drop out the pair of conjugate dynamical variables A_0, B_0 from our scheme.

The theory so obtained is still gauge-invariant since the transformations

$$A_\mu \longrightarrow A_\mu + \frac{\partial S}{\partial x_\mu}$$

are allowed, with S general. Considering one instant of time, S and $\frac{\partial S}{\partial x_0}$ are independent functions of x_1, x_2, x_3. The first gives the gauge transformation

$$A \longrightarrow A^r \frac{\partial S}{\partial x_r}$$

which is produced by the term $\int \kappa B_r{}^r d^3u$ in H new, with $K = S$, and the second gives the gauge transformation

$$A^0 \to A^0 + \frac{\partial S}{\partial x_0}$$

which merely says that A^0 can vary arbitrarily, so that A^0 has no physical importance.

The Hamiltonians, weakly equal to zero in the classical theory, are replaced by wave equations in the Quantum Theory:

$$\left\{ W + \int \left(\tfrac{1}{4} F_{\gamma s} F^{\gamma s} - \tfrac{1}{2} B_\gamma B^\gamma \right) d^3u \right\} \Psi = 0 ,$$
$$B_\gamma{}^\gamma \Psi = 0 .$$

The first of these equations, is a Schrödinger equation and the second a 'supplementary condition' in the usual terminology. But in the present theory both are on the same footing.

For getting a suitable representation, consider the Fourier components:

$$A_\gamma(u) = \int \left\{ A_{k\gamma} e^{-i(\underline{k} \cdot \vec{r})} + \bar{A}_{k\gamma} e^{i(\underline{k} \cdot \underline{u})} \right\} d^3k$$

$$= \int \left\{ A_{k\gamma} + \bar{A}_{-k\gamma} \right\} e^{-i(\underline{k} \cdot \underline{u})} d^3k$$

$$B_\gamma(u) = -i \int |k| \left\{ A_{k\gamma} - \bar{A}_{-k\gamma} \right\} e^{-i(\underline{k} \cdot \underline{u})} d^3k .$$

These decompositions are analogous to the ones we adopted in the scalar case; and we get similarly,

$$[A_{k\gamma}, \bar{A}_{k's}] = - \frac{i g_{\gamma s}}{8\pi^3 |k|} \delta_3 (k - k')$$

where the minus sign on the right is chosen to suit the definition of $g_{\gamma s}$ that we have adopted.

If we take k_γ in the direction of x_3 , A_{k3} gives the longitudinal part and, A_{k1}, A_{k2} give the transverse part of the field .

In general, $\frac{k^r A_{kr}}{|k|}$ gives the longitudinal component: there is no suitable notation for the transverse components; so we introduce a new notation and call then A_{kt}, with t taking on two values 1 and 2.

We have

$$B_r^{\ r} = -\int |k| \, k^r (A_{kr} - \bar{A}_{-kr}) e^{-i\underline{k}\cdot\underline{u}} d^3k.$$

and so the condition

$$B_r^{\ r} \approx 0$$

gives

$$k^r (A_{kr} - \bar{A}_{-kr}) \approx 0$$

Hence the 'supplementary conditions' refer only to the longitudinal components and not to the transverse components.

We now take a standard ket $|S\rangle$ corresponding to the state in which no photons are present. Then $|S\rangle$ satisfies the relations

$$k^r (A_{kr} - \bar{A}_{-kr}) |S\rangle = 0$$

$$\bar{A}_{kt} |S\rangle = 0, \qquad t = 1, 2.$$

Here the first relation refers to the supplementary condition which must be satisfied by any physical state; and the second is the expression of the requirement of no photons being present in the state represented by the ket $|S\rangle$.

A general ket $|P\rangle$ may be expressed in the form

$$|P\rangle = \Psi |S\rangle$$

and to be a physical state, it must satisfy

$$k^r (A_{kr} - \bar{A}_{-kr}) \Psi |S\rangle = 0$$

Taken in conjunction with the relations satisfied by the standard ket, this relation yields

$$[k^\nu (A_{k\nu} - \bar{A}_{-k\nu}) \,,\, \Psi] \; |S\rangle = 0 \, .$$

This condition is satisfied by taking Ψ to be a function of the transverse Fourier components independent of the longitudinal ones. Hence

$$|P\rangle \;=\; (\text{power series in } A_{kt}) \, |S\rangle$$

To obtain the correct Hamiltonian in the quantum theory, one should express B and F_{rs} in terms of the Fourier components (they will involve transverse components only) and rearrange the factors to have no zero point energy. Such a development is straightforward and will yield the Quantum Theory of the Electro-magnetic field when no electrons are present.

To develop the theory of the electromagnetic field when electrons are present, we start with a new action density

$$\mathcal{L} \;=\; \bar{\Psi}\, \alpha_\mu \, (i\psi^\mu + e A^\mu \psi) + m \Psi \alpha_m \psi - \tfrac{1}{4} F_{\mu\nu} F^{\mu\nu} ,$$

where ψ is a new field variable* with 4 components; α_μ are 4×4 matrices occurring in the relativistic theory of the electron**

---

* In this section, ψ will denote a 4-component field quantity (the spinor wavefunction of the electron) and should not be confused with the Ψ used in the previous section.

** Note that the α's are constant matrices and not field quantities.

(the Dirac matrices) with elements $\alpha_{\mu a b}$. According to our general notation

$$\psi^{\mu} = \frac{\partial \psi}{\partial x_{\mu}}$$

$\bar{\psi}\alpha_{\mu}\psi^{\mu}$ for example denotes $\bar{\psi}_a \alpha_{\mu ab} \psi^{\mu}_b$

The matrices α_{μ} have the following properties:

$\alpha_0 = 1,$ $\alpha_1, \alpha_2, \alpha_3, \alpha_m$ anticommute .

The α 's all have squares unity and are Hermitian.

It may be noticed that the action density is not real:

$$\bar{\mathcal{L}} = -i\bar{\psi}^{\mu}\alpha_{\mu}\psi + eA^{\mu}\bar{\psi}\alpha_{\mu}\psi + m\bar{\psi}\alpha_m\psi - \frac{1}{4}F_{\mu\nu}F^{\mu\nu} .$$

Thus,

$$\mathcal{L} - \bar{\mathcal{L}} = i(\psi\alpha_{\mu}\bar{\psi}^{\mu} + \bar{\psi}^{\mu}\alpha_{\mu}\psi) = i(\psi\alpha_{\mu}\bar{\psi})^{\mu}$$

so that,

$$\int (\mathcal{L} - \bar{\mathcal{L}})\, d^4x = 0$$

The pure imaginary part does not contribute anything to the action integral. We can, of course work with the real part $\frac{1}{2}(\mathcal{L} + \bar{\mathcal{L}})$ instead but it is more convenient to work with $\mathcal{L}$ itself.

If χ is any field quantity appearing together with its first derivatives in the action density, then the action principle

$$\delta \int \mathcal{L}\, d^4x = 0$$

leads to the field equation

$$\frac{\partial \mathcal{L}}{\partial \chi} - \left(\frac{\partial \mathcal{L}}{\partial \chi^{\mu}}\right)^{\mu} = 0 ,$$

In the present case, since the real and imaginary parts of ψ_a are independent, we may consider the variations in ψ_a and $\bar{\psi}_a$ to be independent. Take

$$\chi = \bar{\psi}_a .$$

The corresponding field equation is

$$\{\alpha_\mu(i\psi^\mu + eA^\mu\psi) + m\alpha_m\psi\}_\alpha \approx 0$$

which is the usual equation for electrons interacting with the electromagnetic field in the relativistic formulation (with $\hbar = 1$).

Take

$$\chi = \psi_a$$

giving the field equation

$$-i(\bar{\psi}\alpha_\mu)^\mu_a + (eA^\mu\bar{\psi}\alpha_\mu + m\bar{\psi}\alpha_m)_a \approx 0$$

which is the conjugate complex equation to the previous one.

Finally, take, $\chi = A^\mu$

obtaining

$$e\bar{\psi}\alpha_\mu\psi + F_{\nu\mu}{}^{\nu} \approx 0$$

These equations can be interpreted to be Maxwell's equations

$$F_{\nu\mu}{}^{\nu} = J_\mu$$

where J_μ is the charge current density, provided we take

$$J_\mu = -e\bar{\psi}\alpha_\mu\psi .$$

This is the charge current density in the relativistic theory of the electron.

We have thus verified that the action density chosen gives, in conjunction with the action principle, the correct field equations. We now proceed to the generalised Hamiltonian formalism. Put

$$\mathcal{L} = \mathcal{L}_E + \mathcal{L}_F \,, \quad L = L_E + L_F$$

where the subscript F refers to the terms coming from $-\frac{1}{4} F_{\mu\nu} F^{\mu\nu}$ and E to the extra part (due to the presence of the electrons). Then

$$L_E = \dot{y}_0 \int \mathcal{L}_E \, d^3u = \int \bar{\psi} \dot{\psi} \, d^3u + \dot{y}_0 \int (\bar{\psi} \alpha_r \psi_{,r} + \bar{\psi} e A^\mu \alpha_\mu \psi + \bar{\psi} \alpha_m \psi) \, d^3u \,.$$

Introduce the quantities

B_μ conjugate to A^μ

χ_a conjugate to ψ_a

$\bar{\chi}_a$ conjugate to $\bar{\psi}_a$

W conjugate to y_0

Then they satisfy the following P.B. relations:

$$[A^\mu B'_\nu] = \delta_{\mu\nu} \delta_3(u - u')$$

$$[\psi_a \chi'_b] = \delta_{ab} \delta_3(u - u')$$

$$[\bar{\psi}_a, \bar{\chi}'_b] = \delta_{ab} \delta_3(u - u')$$

$$[W, y_0] = 1,$$

and all the other Poisson brackets vanish. We have, for the variation of the Lagrangian

$$\delta L = W \delta \dot{y}_0 + \int \{ B_\mu \delta \dot{A}_\mu + \chi_a \delta \dot{\psi}_a + \bar{\chi}_b \delta \dot{\bar{\psi}}_b \} \, d^3u;$$

computing the variation of the Lagrangian adopted above and comparing coefficients of the elementary variations, we obtain the following weak equations:

$$\chi_a \approx i\bar{\psi}_a$$

$$\bar{\chi}_a \approx 0$$

$$B_0 \approx 0$$

$$B_r \approx F_{r0}$$

$$W \approx \int \{ \bar{\psi} e A_0 \psi + \bar{\psi} \alpha_r (i\psi^r + e A^r \psi) + m \bar{\psi} \alpha_m \psi - \frac{1}{4} F_{rs} F^{rs} + \frac{1}{2} B_r B^r + A_0 B_r{}^{r} \} d^3u$$

From these we have the ϕ equations

$$\chi_a - i \bar{\psi}_a \approx 0$$

$$\bar{\chi}_a \approx 0$$

But the P.b. of these ϕ 's do not vanish

$$[\chi_a - i \bar{\psi}_a , \bar{\chi}'_a] \neq 0$$

and hence the assumption of zero Poisson brackets between the ϕ 's in the general theory is no longer valid; and some new technique* is to be employed.

In the present case such a modification can be done fairly simply. To illustrate the method, take a case in which the ϕ-equations are

$$p_1 - q_2 \approx 0$$

$$p_2 \approx 0$$

We may then use* the equations $p_2 = 0$, $q_2 = p_1$ as strong equations to eliminate q_2, p_2 from the theory. The P.b.'s

---

* For the treatment of the general case see P.A.M.Dirac; Can.Jnl.Math. 2, 129 (1950).

will now be defined** by

$$[\xi, \eta] = \sum_{n \neq 2} \left(\frac{\partial \xi}{\partial q_n} \frac{\partial \eta}{\partial p_n} - \frac{\partial \xi}{\partial p_n} \frac{\partial \eta}{\partial q_n} \right)$$

By this method, we eliminate the $\bar{\psi}, \bar{\chi}$ degree of freedom and work only with the ψ , χ viariables.

Instead, however, we may work with ψ, $\bar{\psi}$ only@ with $i\bar{\psi}$ replacing χ. Then $\bar{\psi}$ with a suitable coefficient is the conjugate to ψ and the commutation relation becomes

$$[\psi_a, \bar{\psi}'_b] = -i\,\delta_{ab}\,\delta_3(u-u') .$$

The general Hamiltonian is

$$\mathcal{H}_{gen} = \dot{y}_0 \{ W - \int \{ \quad \} d^3u + \int \lambda B_0 d^3u$$

The last term in it causes a change of A_0 without change of the surface.

From the field equation

$$e\bar{\psi}\,\alpha_\mu \psi + F_{\nu\mu}^{\ \ \nu} \approx 0$$

we get, for $\mu = 0$

** The following remarks may be made :

1) This definition is invariant under canonical transformations, provided they do not involve p_2 , q_2 . This is not an effective restriction, since p_2 , q_2 no longer appear in the theory.

2) If we go back, from the Hamiltonian so obtained, to the Lagrangian, the Lagrangian obtained need not be identical with the one we started with. However it will be equivalent to the original Lagrangian in the sense of giving the same equations of motion. (We have seen above such a case of two different action densities $\mathcal{L}$ and $\frac{1}{2}(\mathcal{L}+\mathcal{L}_0)$ giving the same field equations).

@ This corresponds, in no sense, to a reintroduction of the $\bar{\psi}$ degree of freedom; and is simply a convenient change of notation.

taking account of the relation

$$B_r^{\ r} \approx F_{r0}^{\ \nu}$$

We must now pass to a new Hamiltonian,

$$\mathcal{H}_{new} = \mathcal{H}_{new} + \int \chi (e\bar{\psi}\psi + B_r^{\ r}) d^3u$$

This Hamiltonian permits changes in gauge of the type

$$A^r \to A^r + \frac{\partial S}{\partial x_r}.$$

with a corresponding change in ψ. These changes are in addition to the changes allowed by the previous Hamiltonian, in which only A_0 is altered.

The terms involving A_0 in $\mathcal{H}_{new}$ are

$$\int A_0 (e\bar{\psi}\psi + B_r^{\ r}) d^3u .$$

These terms can be absorbed in $\int \chi (e\bar{\psi}\psi + B_r^{\ r}) d^3u$ by a different choice of χ. Then A_0 disappears from $\mathcal{H}$. Further, B_0 occurs only in the term $\int \lambda B_0 d^3u$, whose only effect on the equations of motion is to give arbitrary changes in A_0. Thus the A_0, B_0 degrees of freedom can be dropped out, as in the theory with no electrons.

Lecture 22

We found in the last lecture that the Hamiltonian for the electron field interacting with the electromagnetic field is

$$\mathcal{H} = \dot{y}_0 (W + H) + \int \chi \, (\bar{\psi}\psi + B_r B^r) \, d^3u \;, \qquad (1)$$

where

$$H = \int \{ -\bar{\psi}\alpha_r (i\psi^r + eA^r\psi) - \bar{\psi} m \alpha_m \psi + \tfrac{1}{4} F_{rs} F^{rs} + \tfrac{1}{2} B_r B^r \} \, d^3u \,. \qquad (2)$$

The first term in (1) has the effect of making the surface advance in time, the second one that of admitting gauge transformations.

To quantize the theory the field quantities have to be regarded as operators, and the electron field quantities have to satisfy anti commutation relations. The classical ψ 's satisfy the Pb. relations

$$[\psi_a, \bar{\psi}'_b] = -i\,\delta_{ab}\,\delta_3(u - u') \,. \qquad (3)$$

This in quantum theory is changed to

$$[\psi_a, \bar{\psi}'_b]_+ = \delta_{ab}\,\delta_3(u - u') \,, \qquad (4)$$

where $[\xi, \eta]_+ = \xi\eta + \eta\xi$. The commutation relations for the electromagnetic potentials are the same as before.

The wave function Ψ now satisfies the equations

$$(W + H)\,\Psi = 0 \,, \qquad (5)$$

$$(e\bar{\psi}\psi + B_r{}^r)\,\Psi = 0 \,. \qquad (6)$$

Equation (6) does not involve variation with respect to time and may be regarded as a supplementary condition.

To get a representation we resolve the fields ψ, A_r into Fourier components. For A_r this is the same as before. To get the Fourier resolution of ψ, we introduce the three vector $\underline{p} = (p_1, p_2, p_3)$

Define

$$p_0 = \sqrt{m^2 + \underline{p}^2} \,. \qquad (7)$$

Then for each $\underline{p}$ there will exist two Fourier components of ψ_a (for each a) corresponding to positive and negative energy states. We call them ψ_{ap+}, ψ_{ap-}, $a = 1, 2, 3, 4$. Then

$$\psi(u) = \int (\psi_{p+} + \psi_{p-})\, e^{i\underline{p}\underline{u}}\, d^3p \,, \qquad (8)$$

determines $(\psi_{p+} + \psi_{p-})$. To make the decomposition unique we require that

$$(p_0 + \alpha^r p_r + \alpha_m m)\,\psi_{p+} = 0 \,, \qquad (9)$$

$$(-p_0 + \alpha^r p_r + \alpha_m m)\,\psi_{p-} = 0 \,. \qquad (10)$$

From (9) and (10) we get

$$\psi_{p+} = \frac{1}{2}\left(1 - \frac{\alpha^r p_r + \alpha_m m}{p_0}\right)(\psi_{p+} + \psi_{p-}) \,, \qquad (11)$$

$$\psi_{p-} = \frac{1}{2}\left(1 + \frac{\alpha^r p_r + \alpha_m m}{p_0}\right)(\psi_{p+} + \psi_{p-}) \,, \qquad (12)$$

as is readily verified.

From (4), we can now find out the anticommutation relations for ψ_{p+}, ψ_{p-}, from which we get their physical interpretation as emission and absorption operators. With suitable numerical coefficients attached to them,

ψ_{p+} is the emission operator for an electron.

$\bar{\psi}_{p+}$ " " absorption " " " "

ψ_{p-} " " " " " " positron.

$\bar{\psi}_{p-}$ " " emission " " " "

We now introduce a standard ket $|S\rangle$ which corresponds to a state in which there are no electrons, positrons or photons, i.e., a state in which there are no photons and no electrons in positive energy state and in which all the negative energy electron states are filled. Then

$$\left.\begin{aligned} &\bar{\psi}_{p+}|S\rangle = 0, \quad \psi_{p-}|S\rangle = 0 \\ &\bar{A}_{kt}|S\rangle = 0, \\ &(e\bar{\psi}\psi + B_r{}^r)|S\rangle = 0 \end{aligned}\right\} \qquad (13)$$

Any other state will now be expressed as

$$\begin{aligned} &(\text{power series in the emission operators})\,|S\rangle \\ &\equiv F\,|S\rangle. \end{aligned} \qquad (14)$$

This power series must, of course, be antisymmetric in the electron-positron operators.

Any physical $|P\rangle$ should satisfy equations (5) and (6). The latter equation gives

$$(\bar{\psi}\psi + B_r B^r)\, F|S\rangle = 0 \quad , \qquad (15)$$

From (13) and (15) we see that (15) is satisfied if

$$[(\bar{\psi}\psi + B_r B^r), F] = 0 \, . \qquad (16)$$

This is a restriction on the power series. So we must find variables which commute with $(\bar{\psi}\psi + B_r B^r)$ and build our power series from them.

We define a quantity ξ to be gauge invariant if

$$[\xi, (\bar{\psi}'\psi' + B'_r{}^{r'})] = 0 \, . \quad (17)$$

Then B_r, F^{rs} are gauge invariant. We now wish to find out functions of ψ which are gauge invariant.

Consider the quantity

$$\psi(u)\, e^{i\int C_r(u,u')\, A^r(u')\, d^3u'} \qquad (18)$$

in which $C_r(u, u')$ are numbers. We obtain

$$[\, \psi_a(u)\, e^{i\int C_r(u,u')\, A^r(u')\, d^3u'} \, , \, e\, \bar{\psi}''_b \psi''_b + B''_s{}^{s''}\,]$$

KKG:dnn
2/8/55

$$= e^{i\int\cdots}\left\{-ie\delta_{ab}\,\delta_3(u-u'')\psi_b'' + \psi_a\, i\int C_r(u,u')\,\delta_3^{r}{}''(u'-u'')\,d^3u'\right\}$$

$$= e^{i\int\cdots}\,\psi_a\left\{-ie\,\delta_3(u-u'') + i\,\frac{\partial C_r(u,u'')}{\partial u_r''}\right\}. \qquad (19)$$

Hence the quantity (18) is gauge invariant if

$$e\,\delta_3(u-u'') = \frac{\partial C_r(u,u'')}{\partial u_r''}, \qquad (20)$$

i.e., if the divergence of $C_r(u,u')$ with respect to the u' variables vanishes everywhere except at $u = u'$, where it is infinite, corresponding to a source function of strength e. The conjugate complex of (18) is then also gauge invariant.

Thus we see that if we use gauge invariant quantities only, then an electron can appear only together with a certain electro-magnetic field due to the exponential factor in (18). This is the coulomb field of the electron. The present theory is thus more physical inasmuch as an electron is always accompanied by its coulomb field; one never encounters "bare" electrons.

Equation (20) does not determine C_r uniquely, since only the divergence is fixed by it. This arbitrariness corresponds to the possibility of adding transverse waves to the coulomb field of an electron.

With F constructed out of gauge invariant quantities, the equation

$$(W+H)\, F\,|S\rangle = 0 \qquad (21)$$

is the only equation remaining to be solved. This is usually tackled by expanding the wave function in powers of e (taken as a dimensionless constant, since $\hbar = c = 1$).

The present treatment differs from the usual treatment due to Fermi, since longitudinal waves are here treated differently.

The solution of equation (21) by the expansion method leads to infinite integrals in higher orders. The special way in which these divergent integrals occur permits finite results to be extracted in an unambiguous manner by the method of renormalization.

The basic idea of the renormalization method is the following. The parameters m, e which occur in the equations may not be the physically observed mass and charge of an electron. Due to the interaction of the two fields some corrections may be brought in, so that the observed mass and charge are $m + \delta m$ and $e + \delta e$ respectively. So one replaces the parameters m, e in the equations by $m - \delta m$, $e - \delta e$, keeping the symbols m, e for the observed mass and charge. It then turns out that all the infinities can be absorbed in $\delta e, \delta m$. This is satisfactory from a practical point of view. But the method lacks logical consistency since δe and δm are infinite. Moreover, although each term of the series expansion can be made finite by this method, there are indications that the series as a whole is divergent.

It is probable that the source of this difficulty is that we are using the wrong Hamiltonian. There is no compelling argument in favour of it and it is worth while trying to find a better one. In the present theory one can give a meaning to an electron without the coulomb field. The quantity ψ by itself refers to such an electron. Probably in a correct theory it should be impossible to conceive of an electron without the accompanying coulomb field.

One possibility in this direction is to regard, classically, an electron as the end of a single Faraday line of force. The electric field in this picture is built up from discrete Faraday lines of force, which are to be treated as physical things, like strings. One has then to develop a dynamics for such a string like structure, and quantize it. The lack of spherical symmetry of this classical model of the electron gets removed by quantization.

In such a theory a bare electron would be inconceivable, since one cannot imagine the end of a piece of string without having the string.

CPSIA information can be obtained
at www.ICGtesting.com
Printed in the USA
BVOW09s1409120118
505167BV00013B/497/P

9 781614 273349